Techniques in Life Science and Biomedicine for the Non-Expert

Series editor
Alexander E. Kalyuzhny, University of Minnesota, Minneapolis, MN, USA

The goal of this series is to provide concise but thorough introductory guides to various scientific techniques, aimed at both the non-expert researcher and novice scientist. Each book will highlight the advantages and limitations of the technique being covered, identify the experiments to which the technique is best suited, and include numerous figures to help better illustrate and explain the technique to the reader. Currently, there is an abundance of books and journals offering various scientific techniques to experts, but these resources, written in technical scientific jargon, can be difficult for the non-expert, whether an experienced scientist from a different discipline or a new researcher, to understand and follow. These techniques, however, may in fact be quite useful to the non-expert due to the interdisciplinary nature of numerous disciplines, and the lack of sufficient comprehensible guides to such techniques can and does slow down research and lead to employing inadequate techniques, resulting in inaccurate data. This series sets out to fill the gap in this much needed scientific resource.

More information about this series at http://www.springer.com/series/13601

Ashutosh Kumar Shukla

ESR Spectroscopy for Life Science Applications: An Introduction

With contributions by Grzegorz Piotr Guzik,
Wacław Stachowicz, Bernadeta Dobosz,
Ryszard Krzyminiewski, Octavian G. Duliu,
Vasile Bercu

 Springer

Ashutosh Kumar Shukla
Department of Physics
Ewing Christian College
Prayagraj, Uttar Pradesh, India

ISSN 2367-1114   ISSN 2367-1122   (electronic)
Techniques in Life Science and Biomedicine for the Non-Expert
ISBN 978-3-030-64200-6   ISBN 978-3-030-64198-6   (eBook)
https://doi.org/10.1007/978-3-030-64198-6

This Springer imprint is published by the registered company Springer Nature Switzerland AG
The registered company address is: Gewerbestrasse 11, 6330 Cham, Switzerland

To my parents

Preface

It is my pleasure to present this collection with set specific goals of the series Techniques in Life Science and Biomedicine for the Non-Expert. This volume with chapter authors along with me as a co-author in each chapter is intended to present the content in a consistent style. This books talks about electron spin resonance spectroscopy and the applications of this method for different life science applications. Though the applications are in many such fields, the selected chapters here focus on healthcare and pharmaceutical science, paleontology and geochronology, and food science.

I am thankful to the expert contributors for taking time out of their busy schedules. It is only their sincere effort that enabled me to present this text before the audience. My sincere thanks are due to the series editor Dr. Alex Kalyuzhny who has indirectly contributed a lot to make this collection as per series expectations.

I sincerely thank Alison Ball, associate editor, Springer (microbiology and immunology), for giving me the opportunity to present this volume. I also thank Ms. Harithashrivarshini, project coordinator, for her support during the publication process. Though I found that it is difficult to be simple when style of presentation is concerned, I hope at the same time that you will enjoy the text and get attracted to ESR applications in different fields.

Prayagraj, India   Ashutosh Kumar Shukla
August 2020

Contents

Chapter 1
Electron Spin Resonance: An Introduction

Magnetic Susceptibility of Substances

The external magnetic field of intensity H induces dipole moment in magnetically active substance. This dipole moment μ is oriented reversely to magnetic field and defined by the following expression:

$$\mu = \kappa_{ind} \, H \tag{1.1}$$

κ_{ind} is proportionality constant and defines magnetic susceptibility specific for substances. Magnetic susceptibility is used to classify the substances as paramagnets, diamagnets, and ferromagnets.

Paramagnets are the substances of positive magnetic susceptibility near to one (≥ 1). Transition metal ions of copper, manganese, or vanadium, for example, stable organic and inorganic radicals in crystals as well as triplet states, are typical representatives of this group of substance. The structural specificity of paramagnets is the content of unpaired, orbital electron(s) making them suitable for the ESR measurements.

Diamagnets are the substances of negative magnetic susceptibility. This group of substances is represented by valence compounds with paired electrons. Consequently, diamagnets remain magnetically neutral and as such not adaptable for ESR measurements.

Ferromagnets are crystalline substances more or less structurally ordered with magnetic moment evoking their natural magnetism without the intervention of external magnetic field. Iron is a typical representative of this group of substances. Magnetic susceptibility of ferromagnets is positive, of course, and remains extremely high around the magnitude of 1000.

© Springer Nature Switzerland AG 2021
A. K. Shukla, *ESR Spectroscopy for Life Science Applications: An Introduction*,
Techniques in Life Science and Biomedicine for the Non-Expert,
https://doi.org/10.1007/978-3-030-64198-6_1

Paramagnetic Properties and Principle of Electron Spin Resonance

The unique properties of paramagnetic substances are derived from their specific molecular configuration lying in the presence of unpaired electron(s). According to quantum mechanical model, the charged free electrons rotate around their own axes. The rotating motion of electron in magnetic field results in the creation of magnetic moment μ named also spin (Fig. 1.1). The rotating motion of electron around its axis occurs as well if a single electron rotates around the orbital of atomic nucleus integrated with paramagnetic molecule. Such molecular electron possesses magnetic moment μ similar to free electron.

H – intensity of external magnetic field.
μ – magnetic moment of rotating electron,
ω – precession of magnetic dipole in H field.

The phenomenon of electron paramagnetic resonance was discovered in 1942 by Russian physicist E. Zavoisky. Paramagnetic resonance is closely related to Zeeman effect consisting of splitting in the external magnetic field of the basic energy level of paramagnetic molecule characterized by spin quantum number S to $2S + 1$ sublevels.

A simple model of paramagnetic properties of substance is a cloud of free electrons between the poles of electromagnet. With no power supply to electromagnet, there is no magnetic field found between the poles (zero-level intensity of magnetic field). Under such circumstances, free electrons of the cloud remain certainly in chaotic fully incidental positions against the poles, while their individual energy load remains identical. When power supply to electromagnet evokes, immediate, parallel, and antiparallel orientation of electron spins against the direction of magnetic field. Thus, electrons remain grouped in two Zeeman levels, lower and higher from zero energy level of electrons.

Fig. 1.1 Schematic presentation of free electron rotating around its axis in the external magnetic field

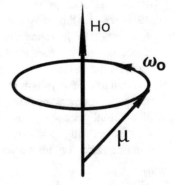

Electron Spin Resonance Phenomenon

The phenomenon of electron spin resonance occurs if paramagnetic substance with unpaired electrons localized in magnetic field and microwaves is incident perpendicular to the direction of magnetic field. The basic resonance condition is expressed by the following formula:

$$\Delta E = g\mu H = h\nu \tag{1.2}$$

where

ΔE is energy difference between the higher and the lower Zeeman levels
H is intensity of external magnetic field.
g is spectroscopic splitting factor,
μ is magnetic dipole moment of electron ($\mu = 9.28477 \times 10^{-24}$ J/T),
h is Planck's constant ($6{,}62607015 \times 10^{-34}$ kg m^{2}s^{-1}),
ν is microwave frequency.

Resonance appears if the linearly increasing magnetic field intensity reaches the value at which the resonant condition expressed in Eq. 1.1 is satisfied followed by the jump of the power carrying by electrons from the lower to higher Zeeman level (Fig. 1.2). The proceeding increase of the intensity of magnetic field disturbs immediately the resonant equilibrium condition and affects the departure of electrons from higher to lower energy level giving rise to the violent absorption of microwave power, called the resonant absorption (Fig. 1.2).

Fig. 1.2 Schematic presentation of resonant absorption. Two-side arrow represents the difference of Zeeman levels under the resonant condition

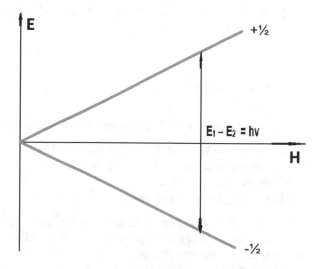

The resonant absorption can be virtually achieved in practice by two ways either by:

1. Gradual increase of the intensity of magnetic field with constant microwave frequency level.
2. Keeping magnetic field intensity constant and increasing of microwave frequency.

In present ESR spectrometers, the resonant absorption is reached by linear increase of magnetic field intensity and keeping of the microwave power frequency constant. The reason lies in technical problem since linear increase of microwave frequency meets difficulty, while the microwave power oscillators of a suitable frequency are available in the market. It is not a problem with the generation of linearly increasing magnetic field with the use of electromagnet, indeed.

The ESR spectrometers currently offered are equipped typically with X-band microwave oscillators emitting microwaves of the frequency 9500 MHz and 3.0 cm wavelength (λ). The resonance condition in X band of organic radicals which are most frequently examined by ESR method today appears at magnetic field intensity 340 ± 10 mT. The wavelength of 30 mm of the microwaves generated in X band makes possible to measure the samples up to 100 mm^3 of the volume. The resolution of single lines in very complex ESR signals in X band is sometimes limited. Therefore, more advanced structural ESR studies the microwave band Q of 35,000 MHz frequency, and wavelength 0.82 cm is adapted and recently even W band of even higher microwave frequency. However, the application of these high-frequency microwave bands is limiting drastically the volume of samples which can be measured. The maximal sample volume in Q band counts 10 mm^3 while that in W band even less, i.e., 0.3 mm^3. The adoption of both Q and W bands in ESR needs special instrumentation. In addition, W-band measurements require extensive cooling of the whole measuring system.

ESR Spectrometer

The basic constituents of ESR spectrometer in X band (see Fig. 1.3) are:

- Electromagnet giving uniform magnetic field between the poles.
- Microwave oscillator–generator like klystron.
- Magnetron or Gunn diode.
- Microwave power crystalline silicon detector.

Electromagnet has precise steering and control system to assure the linear increase of the intensity of magnetic field in time. Microwave power is propagated inside spectrometer components with waveguide to assure minimal energy losses. In X band, the rectangular waveguides are 3.0 cm × 1.5 cm in cross section. In the beginning, the microwave power is conveyed to resonant cavity (resonator) installed centrally between the poles of electromagnet. Two types of resonant cavities are

Fig. 1.3 Rectangular (**a**) and cylindrical (**b**) resonant cavities. The distribution of magnetic component of electromagnetic field inside is shown. Cylindrical cavity demonstrates the entrance of microwave power inside (microwave guide) and localization of sample inside in red

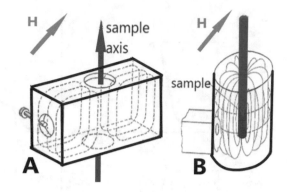

often selected for ESR measurements in X-band rectangular cavity of TE 102 mode and cylindrical cavity of TE 011 mode. Both of them are shown in Fig. 1.3. The sample to be investigated is placed inside the cavities in their symmetry center. The shape and construction of resonant cavity assure multiple and effective reflection of microwaves from the walls of cavities and maximal concentration of magnetic component of electromagnetic field in the symmetry center of resonator.

The more multiple the stationary distribution inside resonance cavity is attained, the higher becomes the sensitivity of EPR measurement involved. In order to improve the efficient reflection of microwaves from cavity walls, the interior is coated with the layer of pure gold or iridium and then excellently polished.

The microwave frequency and magnetic field intensity are registered and automatically stored in the memory of spectrometer layer computer.

The quality of resonant cavities is defined by Q factor which expresses the ratio of the volume of microwave energy cumulated inside the cavity to the losses of this energy dispersed in cavity walls and in holes used for insertion of sample tubes to the cavity. Q factor of commercially produced resonant cavities is usually between 3000 and 10,000. The waveguide alongside two devices are installed. The first one is ferrite insulator which prevents the system against accidental back flow of microwaves inside waveguide system, while the second is microwave absorber for the adjustment of microwave power needed.

The microwave power flow carrying the signal of resonance absorption reaches the microwave detector where it becomes registered and automatically amplified 10^5 to 10^6 times and converted to the first derivative form and recorded.

First Derivative of the ESR Signal

The ESR signal of investigated solid or liquid sample if registered in the primary Gaussian or Lorentz forms depending on its molecular neighborhood is relatively broad despite whether composed of a single (Fig. 1.4a, the upper signal) or several spectral lines. In consequence, the resolution of single lines in multiline signal

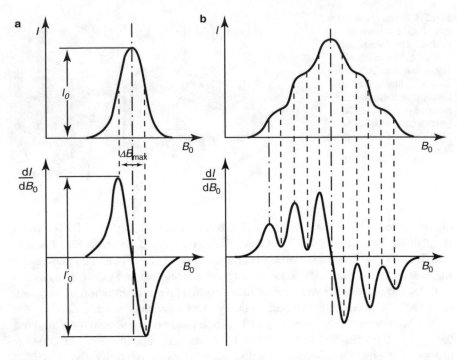

Fig. 1.4 Basic (upper signals) and first derivative (lower signals) ESR signals of DPPH standard sample in the crystalline form (**a**) and dissolved in benzene (**b**). Better resolution of ESR lines recorded in the form of first derive is obvious. (Reproduced from Radiation sterilization of tissue grafts, EU/INCT Project Edition; Warszawa 2009; page 160, Fig. 10. In Polish only)

does not remain satisfactory for further spectroscopic analysis (Fig. 1.4b, the upper signal). The solution was found in the transformation of basic ESR signal to the form of the first derivatives shown in Fig. 1.4a, b, lower signal. Presently, the presentation of ESR signals in the form of first derivative is absolutely common. The basic signals of resonant absorption are automatically transformed to the first derivative signal in the ESR spectrometers. This presents now well-resolved sharp peaks (Fig. 1.4, lower signals a and b). The primary resonant absorption signals as presented in Fig. 1.4, higher signals a and b, are no more available and can be obtained optionally with the use of more sophisticated ESR spectrometers only. The basic and first derivative ESR signals shown in Fig. 1.4 were obtained with powdered (single line) and dissolved (five lines) DPPH (diphenyl-picryl-hydrasil), the stable organic radical used in ESR as a popular standard.

Hyperfine Splitting

Hyperfine splitting of the ESR signals appears if unpaired electron of paramagnetic substance interacts with one or more nuclei with non-zero electron spin of atoms belonging to the same paramagnetic molecule. Despite of external magnetic field, the unpaired electron interacts with a weak magnetic field of the nuclei of atoms belonging to the same paramagnetic molecule. Let us consider the case of hyperfine interaction of unpaired electron with magnetic field of the nucleus of one hydrogen atom. Nuclear spin of hydrogen is $\pm \frac{1}{2}$; thus, two orientations of spins are allowed, parallel and antiparallel to the magnetic field. Magnetic moment of hydrogen atom which is 10^3 times lower than that of unpaired electron creates a weak magnetic field in its vicinity of unpaired electron. This weak magnetic field slightly increases and from the other side decreases the magnetic field of electron which means that each of two electron levels ($+ \frac{1}{2}$ and $- \frac{1}{2}$) undergoes splitting in two sublevels. Under the resonant condition, the splitting of a single signal of electron in two component lines occurs. The effect is called hyperfine splitting (Fig. 1.5).

The intensity of magnetic field at which resonant absorption of microwave power occurs is expressed with two equations:

$$H_1 = h\nu/g\,\mu - \Delta H \tag{1.3a}$$

and

$$H_2 = h\nu/g\,\mu + \Delta H \tag{1.3b}$$

Fig. 1.5 Hyperfine interaction of unpaired electron with the nucleus of one hydrogen atom with spin resulting in the splitting of primary signal in two sublevels giving rise to isotropic doublet distanced A between two lines

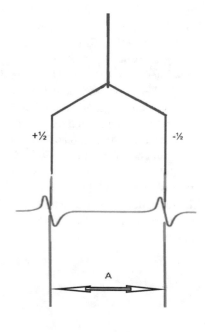

Table 1.1 The number of ESR lines and proportion between their intensities depending on the number of equivalent H atom nuclei (spin ± ½) interacting with unpaired electron

Number of H atom nuclei interacting with unpaired electron	Number and intensity of ESR lines obtained	Type of ESR signal obtained
0	1	Singlet
1	1:1	Doublet
2	1:2:1	Triplet
3	1:3:3:1	Quartet
4	1:4:6:4:1	Quintet
5	1:5:10:10:5:1	Sextet
6	1:6:15:20:15:6:1	Septet

where

$$\Delta H = H_1 - H_2 = A/\mu g \qquad (1.3c)$$

A is the hyperfine splitting factor.

Hyperfine splitting factor A characterizes the interaction between nuclear spins of atoms and unpaired electron. In the case of atomic hydrogen, it accounts 50.7 mT, for example. Under the equally strong interaction of unpaired electron with more than two nuclei, the effect of the overlapping of spectral components of resonant absorption takes place since all sublevels are equally distanced by ΔH. In consequence, symmetric signal is obtained composed of $2s + 1$ spectral lines with maximal intensity at the central lines. The number of lines and the distribution of line intensities depending on the number of interacting nuclei are along with Pascal triangle mathematical model given in Table 1.1.

In other words, if the interaction of unpaired electron with two hydrogen nuclei having the same spin equal to ± ½ occurs, the symmetric signal composed of three lines with the intensity distribution 1:2:1 is obtained. Similarly, interaction of unpaired electron with three equivalent hydrogen nuclei, in turn results the splitting of basic signal to four sublevels, occurs, and the ESR signal is composed of four lines representing the intensity distribution 1:3:3:1, respectively (Table 1.1).

There are also situations known if in the consequence of specific composition and configuration of paramagnetic molecule unpaired electron interacts with nitrogen atom(s) or with hydrogen and nitrogen nuclei of spin equal to ±1. Such situation leads obviously to the formation of different and usually complex ESR spectra. The formation of very different multiline ESR signals by searching of paramagnetic substances is a basis for further studies on the elucidation of radical structures, for example.

Spectroscopic Splitting Factor g

Spectroscopic splitting factor g characterizes the magnetic moment of any paramagnetic molecule and represents the number specific for a given paramagnetic substance studied. The numerical value of g factor is obtained by transforming of the basic resonance Eq. (1.2) resulting in

$$g = h\nu/\mu H \qquad (1.4)$$

where

h – Planck's constant – 662607×10^{-14} J*s,
ν – microwave frequency,
H – the intensity of magnetic field signed also as B.
μ – Bohr magnetron – $9{,}2732 \times 10^{-21}$ erg/gauss.

The g factor of free electron with the relativistic correction equals to 2.00229. For the most of organic radicals with unpaired electron interacting with the nuclei of carbon, oxygen, sulfur, and hydrogen, g factor differs only slightly from that of free electron and equals typically from 2.002 to 2.004 depending on state of the delocalization of electrons inside its molecules. For that reason, g factor remains an important numerical value which describes any substance investigated by the ESR spectroscopy.

In the above Eq. (1.4) h and μ are constants while H is a variable only. The microwave frequency ν is virtually stable if ESR measurements are conducted at X band only. Microwave frequency is obviously changed if the other frequency bands like Q or W are used. Certain differences of this parameter can also appear, if the results are obtained with the use of different ESR spectrometers equipped with the other types of microwave generators. The precise estimation of g factor is sometimes difficult especially if complex, anisotropic signals of powdered paramagnetic substances are concerned.

Spin–Orbital Interaction and Anisotropy of the ESR Signal

Spectroscopic splitting factor g from the basic equation of paramagnetic resonance (1.1) is connected with spin–orbital interaction of unpaired electron which depends on the state of investigated substance. In gaseous and liquid samples including some amorphous substances, spin–orbital interaction is negligible or low and g factor is at the same mean level. However, if radical is incorporated strongly with the crystalline lattice of the substance, the orbital motion becomes limited, while orbital moment does not undergo free orientation and spin–orbital coupling is changed. The

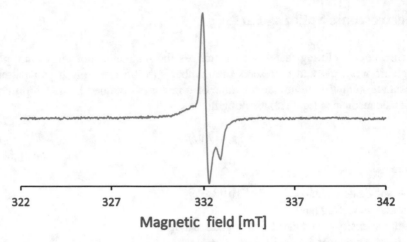

Fig. 1.6 The ESR signal of irradiated bone sample containing crystalline hydroxyapatite of axial symmetry characterized by $\mathbf{g}_\parallel = 1.997$ and $\mathbf{g} = 2.003$, respectively

consequent anisotropy of g factor is followed by the deformation and simplification of ESR signals involved. The effect becomes very well observed with mono crystals by rotation of crystal sample inside resonant cavity, when g values are taking different values by positioning alongside with crystal symmetry axes. Anisotropic ESR signals of polycrystalline samples are asymmetric and strongly deformed and characterized by two or three g factors depending on crystal structure. As an example, the asymmetric ESR singlet obtained with irradiated bone containing crystalline hydroxyapatite is shown in Fig. 1.6.

Line Width

Theoretically, the line width of ESR signals should be negligible since transmission of electrons from higher to lower Zeeman level occurring by resonant condition is extremely quick. However, the ESR signals recorded in practice are markedly broadened and giving rise to Gaussian or Lorenz peaks of smaller or bigger line widths which is caused by the influence of internal and external factors. The internal factors influencing the line width of ESR signals are independent on measuring condition and result from the interaction of spins with surrounding atoms and chaotic movements of molecules in any real multi-molecular system.

The external factors play essential role on the shape and line width of the ESR signals recorded which depend virtually on the condition how the ESR measurement is performed. Therefore, the mutual optimization of measuring parameters is of great importance indeed. If measuring parameters are not properly chosen, this results in the deformation and uncontrolled broadening of the ESR signals involved.

Special attention has to be paid whether the adjustment of three basic parameters of the ESR measurement is correctly selected. These are:

- Microwave power [mW].
- Amplitude of modulation [G].
- Time constant [s] adjusted for the recording of investigated signal.

Spin Saturation Effect

The higher the level of microwave power used, the more convenient becomes the condition of the recording of the ESR signal involved (low noise level, more intense ESR signals, etc.). However, slow relaxation processes which occur in the investigated systems and limit the rate of filling Zeeman energy levels evoke the disadvantageous phenomenon of spin saturation which usually appears at higher microwave power adjustment. The population of spins on both lower and higher Zeeman energy levels became alike and resonant absorption anneals. The effect is reflected in the diminishing and broadening of the ESR signal which sometimes disappears as well. For that reason analysis of a new sample of unknown properties for the first time, it is recommended to proceed the control analysis in advance with the intention to estimate the critical maximum level of microwave power adjustment. The measurement on unknown sample ought to be executed at microwave power level lower than that estimated in control analysis. Figure 1.7 demonstrates the ESR recordings of a sample with and without proper adjustment of microwave power.

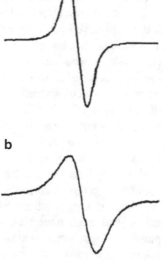

Fig. 1.7 The example of the ESR signals recorded with (**a**) and without (**b**) properly adjusted microwave power. (From Stachowicz W., Identyfikacja napromieniowanej żywności, Postępy Techniki Jądrowej. Biblioteka Nukleoniczna; Warsaw 2005; vol. 48, z.4; pp. 25–32, in Polish only)

Amplitude of Modulation

In order to obtain the recording of resonant absorption in the form of the first derivative, the primary ESR signal undergoes modulation. Most frequently, 100 kHz modulating current is applied and delivered to resonant cavity by means of the coil which is built inside electromagnet poles. The amplitude of modulation applied routinely should be relatively low to prevent any disturbance of the shape of the first derivative ESR signal obtained. It is because by the increasing of modulation amplitude, the broadening of the ESR signal appears which limits the resolution of single lines if complex, multiline signals are concerned. The evident broadening of the signal appears if the amplitude of modulation is approximately equal to the half of the width of a single spectral line. Therefore, in order to prevent the broadening of ESR signal and to obtain optimal result, it is recommended to adjust modulation amplitude on the level approximately equal to one tenth of the corresponding half width of the signal involved. On the other side, however, the higher modulation amplitude is adapted, the better quality of the ESR signal (higher intensity and low signal-to-noise level) is obtained. Thus, for the study of very weak ESR signals, the application of relatively high amplitude modulation is still recommended.

Applications

ESR spectroscopy is widely used in different branches of science like physics, chemistry, biochemistry, biology, archaeology, geology, medicine, and food science. Paramagnetic defects in crystals and those generated by ionizing radiation, electrons in conduction layers, semiconductors, and magnetic susceptibility studies are a few to mention as applications in physics.

Molecular structures of a lot of paramagnetic substances of chemical or biochemical origin have been discovered by means of ESR spectroscopy-based methods. The ESR studies on transient compounds produced in catalytic processes lead to important information essential for the understanding of the responsible mechanisms. Crucial information on the structures of transient chemical species and their behavior has been studied using low-temperature ESR studies. The identification of radicals stabilized at liquid nitrogen temperature in deep frozen matrices makes possible to undertake kinetic studies on their transformation and decay. Special attention has been paid to the studies on long-lived and short-lived radicals produced by means of ionizing radiation and their further reactions. ESR methods were found useful in studies on organ–metallic compounds containing paramagnetic transition metal ions, redox chemical reactions, and polymerization processes. Kinetics and structure of stable radicals and ions/radicals produced by ionizing radiation in tissues and foods have been studied using ESR. Mechanism of photosynthesis in plants has been better understood applying ESR methods. Relationship between the number of unpaired electrons and the intensity of light stimulating photosynthesis has also been studied. The ESR measurements are successfully employed in dosimetry for the identification of food irradiation as well as in archaeology for the estimation of the age of osseous tissues survived in animals and human bones.

Following chapters will give you a glimpse of applications in different fields of interest.

References

Ayscough P.B.: Electron Spin Resonance in Chemistry. Van Nostrand Reinhold Co. New York, Toronto, London, Melbourne, 1967;

Hedvig P., Zantai G. Microwave study of chemical changes and reactions, AkademiaiKiado, Budapest, 1967;

Symons M., Electron spin resonance spectroscopy, Van Nostrand Reinhold Co, 1978;

Kęcki Z., Principles of molecular spectroscopy, PWN Warsaw, 1992, In Polish only;

Ikea M. New applications of Electron Spin Resonance – Datong Dosimetry and Microscopy, World Science Publishing Co., Singapore, New Jersey, London, Hong Kong 1993, 502 pp;

Guzik G.P., Stachowicz W.: The measurement of the luminescence stimulated with the light, the quick method of the identification of radiating of the food; INCT Report 2005. Serie B No. 3/2005; 16 pp, In Polish only;

Guzik G.P., Stachowicz W., Michalik J.: Study on stable radicals produced by ionizing radiation in dried fruits and related sugars by electron paramagnetic resonance spectrometry and photostimulated luminescence method – I.D-fructose; Nukleonika 2008; vol.53: S89–S94.

Guzik G.P., Stachowicz W., Michalik J.: Study on organic radicals giving rise to multicomponent EMR spectra in dried fruits exposed to ionizing radiation; Current Topics in Biophysics 2010; vol.33 (suppl. A): 81–85.

Guzik G.P., Stachowicz W., Michalik J.: EPR study on sugar radicals utilized for detection of radiation treatment of food; Nukleonika 2012; vol.57 (4): 545–549.

Kurreck H., Elger G., Von Gersdorff J., Wiehe A., Mobius K.: EPR studies of photoinduced electron transfer in triad model compounds of photosynthesis, Applied Magnetic Resonance vol.14, issue 2–3, 203–215 (1998).

Kroh J.: Free radicals in radiational chemistry; the Scientific Publishing House PWN, Warsaw 1967; In Polish only;

Scientific Editor J. Michalik, Radiation sterilization of tissue grafts, EU/INCT Project Edition, Warszawa 2009; 135–224, In Polish only;

Moens P., Callens F., Matthys P., Maes F., Verbeek R., Naessens D.(1991): Adsorption of Carbonate-derived molecules on surface of carbonate-containing apatites: J. Chem. Soc. Faraday Trans. 87, 3137–3141.

European Standard PN-EN 13708:2003: Foodstuffs: Detection of irradiated food containing crystalline sugar by ESR spectroscopy,. European Committee for Standardisation (CEN), Brussels, PKN Polish version;

European Standard PN-EN 13751: 2009: Foodstuffs: Detection of irradiated food by photoluminescence, European Committee for Standardisation, (CEN), Brussels, PKN Polish version,

European Standard PN-EN 1784,Foodstuffs – Detection of irradiated food containing fat – Gas chromatographic analysis of hydrocarbons, European Committee for Standardisation (CEN) Brussels, PKN Polish version;

European Standard PN-EN 1785, Foodstuffs -Detection of irradiated food containing fat – Gas chromatographic/mass spectrometric analysis of 2-alky/cyclobutanones, European Committee for Standardisation (CEN), Brussels, PKN Polish version;

European Standard PN-EN 1786:2000: Foodstuffs -Detection of irradiated food containing bone –
    Method by ESR spectroscopy. European Committee for Standardisation (CEN), Brussels PKN
    Polish version;
European Standard PN-EN 1787:2001: Foodstuffs – Detection of irradiated food containing
    cellulose by ESR spectroscopy. European Committee for Standardisation (CEN), Brussels,,
    PKN Polish version;
European Standard PN-EN 1788:2002: Foodstuffs – Thermoluminescence detection of irradiated
    food from which silicate minerals can be isolated, European Committee for Standardisation,
    Brussels, PKN Polish version;
Sobkowski J.: Radiation chemistry and radiation safety; The Publish House Adamantan, Warsaw
    2009; in Polish only.
Stachowicz W., Strzelczak G., Michalik J., Wojtowicz A., Dziedzic-Gocławska A., Ostrowski K.
    Application of EPR Spectroscopy for Control of Irradiated Food; J. Sci. Food Agric. 1992; vol.
    58, 407–415.
Stachowicz W., Burlińska G., Michalik J., Dziedzic-Gocławska A., Ostrowski K.: Applications of
    EPR Spectroscopy to Radiation Treated Materials in Medicine, Dosimetry, and Agriculture, in:
    Appl. Radiat. Isot.1993a, vol.44, No 1/2, 423–427.
Stachowicz W., Burlińska G., Michalik J., Dziedzic-Gocławska A., Ostrowski K.: The role of
    long lived EPR active species produced by ionizing radiation in biological materials as used
    in medical, dosimetric and agricultural studies; Nukleonika 1993b; vol.38, No 3, 67–82.
Stachowicz W., Burlińska G., Michalik J., Dziedzic-Gocławska A., Ostrowski K.: The EPR
    detection of foods preserved with the use of ionizing radiation. Radiation Physical Chemistry
    1995, vol. 46, no.4–6, 771–777,
Stachowicz W., Burlińska G., Michalik J., Dziedzic-Gocławska A., Ostrowski K.: EPR spec-
    troscopy for the detection of foods treated with ionizing radiation; in: Detection methods for
    irradiated foods. Current status, Special Publication No.171. The Royal Society of Chemistry
    1996, pp.23–32
Stachowicz W., Identyfikacja napromieniowanej żywności, Postępy Techniki Jądrowej. Biblioteka
    Nukleoniczna; Warsaw 2005; vol. 48, z.4; pp.25–32,
Stankowski J., Hilczer W.: Introduction to magnetic resonance spectroscopy, PWN Edition,
    Warsaw 2005, In Polish only,
Zagórski Z.P.: Radiation sterilization, INCT Edition, Warsaw 2007, In Polish only,
Zavoisky E.: Spin magnetic resonance in paramagnetic substances; the Journal of Physics of the
    USSR, Moscow 1945, 92, 45-55.
EMR/ESR/EPR spectroscopy for characterization of nanomaterials, Ashutosh Kumar Shukla (Ed.)
    A part of Springer series on Advanced Structured Materials, Vol. 62 (2017), eBook ISBN 978-
    81-322-3655-9, Hardcover ISBN 978-81-322-3653-5, Series ISSN 1869-8433
Electron Spin Resonance Spectroscopy in Food Science, Ashutosh Kumar Shukla (Ed.), Elsevier
    (2017), Paperback ISBN: 9780128054284
Electron Spin Resonance in Medicine, Ashutosh Kumar Shukla (Ed.), Springer(2019) Hardcover
    ISBN 978-981-13-2229-7, eBook ISBN 978-981-13-2230-3
Electron Magnetic Resonance – Applications in Physical Sciences and Biology (A part of Elsevier
    series entitled Experimental Methods in Physical Sciences) (2019), Ashutosh Kumar Shukla
    (Ed.), Paperback ISBN 9780128140246, eBook ISBN: 9780128140253

Chapter 2
ESR: Applications in Healthcare and Pharmaceutical Science

Sterilization

Sterilization is defined as a process which eliminates (kills, removes, or deactivates) all forms of life and biological elements (e.g., bacteria, viruses, etc.) present in a specific area (surface, volume, culture media, etc.).

In the Middle Ages, medical equipment was sterilized by annealing surgical instruments on fire. Currently, there are various methods which can be used for this purpose, the choice of which depends on the material that we want to sterilize. They can be divided in two groups. One of them are physical methods, including thermal (or heat) methods, high pressure, ultrasonic vibrations, filtration, and irradiation. To the second group, chemical methods, belong gaseous methods: ethylene oxide, formaldehyde, ozone and plasma. If we would like to divide physical methods more precisely, we can create two groups: methods which use heat (flaming, incineration, dry heat, and autoclave steam sterilization) and non-heat sterilization methods.

Sterilization has a potential application in many fields such as food, medicine, pharmacy, and spacecraft. At this moment, the most important for this chapter are healthcare (medicine) and pharmacy. The most popular methods are here: moist and dry sterilization, gaseous sterilization, and irradiation and filtration sterilization.

Brief Description of Sterilization Methods

The most popular method of sterilization is heating which involves the destruction of enzymes and other cells structures.

© Springer Nature Switzerland AG 2021
A. K. Shukla, *ESR Spectroscopy for Life Science Applications: An Introduction*,
Techniques in Life Science and Biomedicine for the Non-Expert,
https://doi.org/10.1007/978-3-030-64198-6_2

Moist Heat Sterilization (121–134 °C)

It takes from 3 to 15 min. The popular method in this group is autoclaving which uses steam under pressure. An effective biocide is used for surgical and diagnostic equipment, surgical dressings, water injections, sheets, containers, ophthalmic preparations, and irrigation fluids.

Dry Heat Sterilization (160–180 °C)

This method requires high temperatures and time of exposure up to 2 h. The method is used for pharmaceuticals that are thermally stable, are impermeable to moisture, and are sensitive to it. This method is applied in case of oils, fatty waxes, oily injections, ointments, and bases for them, dry medications in the form of powders, and medications in the form of non-aqueous suspensions.

High Pressure

This kind of sterilization is mainly used for food. Under the influence of high pressure, the number of microorganisms decreases significantly. In this way, the shelf life of the product is extended. The research is conducted on the impact of high pressure on microorganisms, including pathogenic bacteria and molds that cause food-borne diseases, on the possibility of destroying bacterial spores and food safety research. Application of high pressure for pasteurization and extension of shelf life of various food products is used for freshly squeezed fruit and vegetable juices, fruit desserts, and meat and fish products.

Ultrasonic Vibrations

They are high-frequency sound waves inaudible to the human ear. In fluids ultrasonic vibrations cause the formation of microscopic bubbles. The cavities rapidly collapse, send out shock waves, and can cause cavitation of microorganisms (bacteria), during which they are quickly disintegrated by the external pressures.

Ultrasonic devices are used for cleaning dental plates, jewelry, and coins. Hospitals use ultrasonic devices to clean their instruments, and research laboratories use ultrasonic probes for cell disruption. In combination with an effective germicide, an ultrasonic device may be used for sterilization; however, other methods are more efficient.

Filtration

The effectiveness of filtration process depends on the material's quality of which the filter element is made and the pore size. Using the new generation of membrane filters made of a mixture of cellulose esters, polyester, nylon, teflon, or fiberglass, the elimination of bacteria, fungi, and viruses is possible. The principle of filtration is based on the under pressure in the filtrate collection container (filter on the flask disconnected to the vacuum pump) or the pressure exerted on the solution to be filtered (syringe with filter cap).

This method is applicable in biological products, heat-sensitive injections and ophthalmic solutions, air, and other gases to be supplied in aseptic areas and in industry as part of venting systems for centrifuges, autoclaves, freezers, and fermenters. In addition, membrane filters are used in sterility tests.

Irradiation

In radiation sterilization UV, X-rays or gamma rays are used. Their effectiveness is connected with different ability of penetration. UV has low penetration and less effectiveness, but at the same time, it is relatively safe. X-rays and gamma irradiation are more dangerous and can penetrate deeper and are more effective as methods of sterilization. For UV sterilization, 260 nm wavelength is usually applied because of the highest bactericidal effectiveness. Sterilization by ionizing radiation is connected with the discovery of X-rays by W.C. Roentgen in 1895, the observation of radiation emitted from uranium by H. Becquerel in 1896, and the discovery of the radioactive elements by Marie and Pierre Curie in 1898. Of course, in those times people did not know the damaging effect of radiation on the human body. The first equipment for sterilization by radiation was installed in 1963 in the USA, and 1 year later, the Atomic Energy of Canada started the first industrial cobalt (^{60}Co) sterilization facility. It was first used to sterilize medicinal substances in the 1980s.

Ionizing radiation is defined as all types of radiation that cause the separation of at least one electron from an atom, molecule, or crystal structure. The sources of such irradiation in industry and pharmacy are most often devices with radioactive isotopes of cobalt and cesium. High-energy electrons with energy not exceeding 10 MeV are used in radiation treatment. The interactions between electromagnetic radiations can be described by three phenomena:

(a) A photoelectric effect (a low-energy photon is absorbed in an atom, and its energy is emitted in the form of a photoelectron; a pair of ionized atom and photoelectron is formed)

(b) The Compton effect (a photon loses some of its energy to electron emissions, and the scattered photon changes the original angle of incidence on a given atom)

(c) Formation of electron-positron pairs (occurs when the photon energy exceeds 1.022 MeV; electrically charged particles are annihilated with the emission of two new photons with energies of 0.511 MeV)

Currently, electron accelerators are used in the radiation sterilization process. This allows shortening the exposure time, easy control of the entire process, and the ability to turn off the electron beam at any time.

This type of sterilization is usually applied to "dry" materials: surgical instruments, prostheses and pharmaceuticals (antibiotics, hormones), sutures, and catheters.

There are many benefits to using this method for medicines such as the lack of contact of the sterilized drugs with chemicals and thus their contamination. Because gamma irradiation is very penetrating, medicines can also be irradiated when packaged. After such sterilization, the products do not show radioactivity. Finally, this method is very effective because of the sensitivity of microorganisms to this radiation.

Microorganisms are much more resistant to radiation than higher organisms. For this reason, different doses of irradiation are applied in sterilization compared to medical treatment (the impact of irradiation on human body is described later). The dose of gamma irradiation applied for medical products should be at least 25 kGy; however, the doses of irradiation are individually chosen by a producer and are in the limits 10–50 kGy. Such doses can change the chemical structure of irradiated product and thus its properties, disqualifying it from the possibility of radiation sterilization. The changes, caused by radiation, in the profile of pharmacological activity in consequence may be dangerous to human health and life. Besides the dose of irradiation, the source of it is also important. The electron beam is most often used due to the documented lower destructive effect.

During this sterilization, free radicals are generated because such radiation breaks chemical bonds. The only direct method for measuring free radicals is electron spin resonance (ESR). From such measurements, the information about both the types of free radicals and about their concentration and dynamics can be obtained. Exemplary ESR spectrum is shown in Fig. 2.1. For each spectrum characteristic spectroscopic parameters are determined: spectroscopic splitting factor (g-value) – characteristic for each paramagnetic species and free radicals, peak-to-peak line width (ΔB), resonance field (B_r), and amplitude of the signal. If the spectrum has a complex structure, hyperfine splitting constant (A) is also determined. There are two types of interactions and ESR spectrum structure. One of them is so-called fine structure, connected with the interaction between unpaired electrons, and the second one, hyperfine structure, connected with the interaction between an unpaired electron and the nucleus spin. To get the information about the concentration of free radicals, the area under the absorption curve should be calculated. To do this, it is necessary to double integrate the ESR signal (first integration gives the absorption curve, the second the concentration).

The undoubted advantage of the ESR method is the possibility of samples measuring in various forms. It is possible to study crystalline drugs, powders, and

Fig. 2.1 Exemplary ESR spectrum: (**a**) recorded from an experiment as the first derivative of absorption curve with marked spectroscopic parameters, (**b**) absorption curve with the concentration of free radicals, (**c**) exemplary ESR spectrum with hyperfine splitting

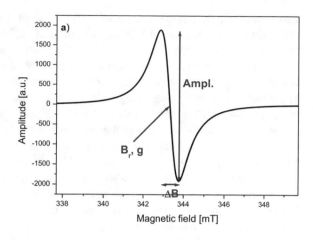

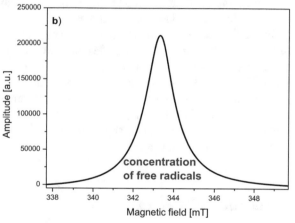

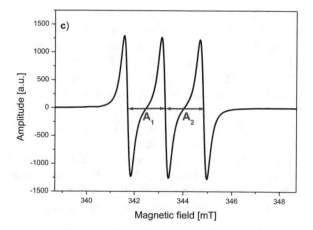

also liquids (suspensions). Quartz or polymer vials are used for measurements, which do not give additional ESR signals. Special capillaries and cuvettes are used to measure the solutions, which allow measurements of samples with water content. In this way, the effects of sterilization can be controlled by the ESR method for sample in any form. Additionally, using temperature control unit, it is possible to measure the sample depending on the temperature and by the way the impact of temperature on the stability of free radicals. The most important is that ESR measurement is not destructive to the sample and by the way it is very precise and sensitive. Briefly, the studied sample is placed in a resonance cavity located in the magnetic field. Microwaves are generated in the microwave bridge and are delivered to resonance cavity with the studied sample using a waveguide. If the range of electromagnetic field we sweep the sample is properly selected for the microwave frequency, we observe the ESR phenomenon and the ESR spectrum on the computer screen. Remember you have studied the basic resonance condition in Chap. 1. The chosen examples of sterilization using ionizing radiation and ESR application are shown in this chapter later, in "Examples."

In addition to sterilization, the radiation method is also used in the polymerization of hydrogels, i.e., gels with a high degree of cross-linking. These materials, thanks to the high water vapor permeability, constitute a new generation of dressings for difficult to heal wounds. Radiation cross-linked polymers are the material for the production of contact lenses or implants replacing natural organs and are also used as matrices for the controlled release of drugs. In addition, the possibility of radiation preparation of metal nanoparticles (e.g., silver) in polymer matrices (gels) gives them bactericidal properties.

Sterilization with Plasma

One of the newer methods of sterilization, which has been popular in recent years, is surface modification with plasma. This method is especially applied to polymeric materials. Plasma is a partially ionized gas (or gas mixture) consisting of approximately equal numbers of electrons, ions, atoms, inert particles, and electromagnetic radiation. All these particles can be found in the ground or excited state. It is used to disinfect surgical instruments and has a biocidal effect on viruses, bacteria, and fungi. Under the influence of plasma, between nitrogen, oxygen, and water vapor, reactions occur in which substances with a strong disinfecting effect are formed. For example, within 12 s, the number of microorganisms on the plasma-treated surface of the hands decreases a million times. In literature, the information can be found that plasma can accelerate wound healing or heal the gums. This technique can be applied to sterilize biodegradable products, as well as in the field of new technologies for preserving or protecting food.

Gaseous Methods

These methods are classified as chemical methods. In this way the problem of heat damage is avoided, but on the other hand, because of their properties, they are usually harmful to people.

The most common used is ethylene oxide treatment (EO, EtO). It is so effective because it penetrates all porous materials including films and some plastic materials. EO kills all known microorganisms (viruses, bacteria, and fungi) and is compatible with almost all materials even when repeatedly applied. Another advantage is its compatibility with most materials and the possibility of multiple uses. It is flammable, toxic, and carcinogenic.

Nitrogen dioxide (NO_2) gas is a second method in this group effective against a wide range of microorganisms (bacteria, viruses, and spores). It causes the degradation of DNA. Liquid NO_2 may be used as the sterilant gas. NO_2 is compatible with most medical materials and less corrosive than other sterilant gases.

Another gas, ozone, is used in industry to sterilize water and air and for surfaces disinfection. Its benefit is the ability to oxidize most organic matter, but on the other hand, it is very dangerous, unstable, and toxic gas that must be produced on-site, so it is not practical to use in many situations. Because of its strong oxidizing properties, it is very efficient and capable of destructing a lot of pathogens.

There are also some chemicals in a form of liquids used in sterilization. Glutaraldehyde and formaldehyde solutions are two of them. They are toxic both for the skin and inhalation. Another sterilizing agent is hydrogen peroxide. Because of strong oxidant properties, it can destroy a lot of pathogens. Hydrogen peroxide is used to sterilize temperature-sensitive devices, such as rigid endoscopes. Peracetic acid (0.2%) is a next sterilant used in sterilizing medical devices such as endoscopes.

Such methods have application in various heat-sensitive drugs, vaccines, hormones, proteins, etc.

The Advantages of Radiation Sterilization of Medicines

Among all the methods used to sterilize drugs, there are features that cause the radiation sterilization method to have an advantage over others. One of them is the reliability resulting from the mechanism of the influence of ionizing radiation on living organisms. Another advantage is the absence of radiation treatment residues in the sterilized drug. The radiation method has the advantage over the gas method, in which desorption of the ethylene oxide used is necessary. At the same time, radiation is more permeable inside the sterilized material than ethylene oxide. In addition, the radiation sterilization process does not harm the natural environment, guaranteeing the storage efficiency of the drug for up to 5 years. Another advantage is the possibility of sterilization in any packaging. The advantage of this sterilization method is also the possibility of carrying out it at any temperature, also below

0 °C. This makes it possible to sterilize drugs that contain highly thermolabile compounds. It is also important that drugs in any form and those with a high level of reactivity can undergo radiation sterilization.

Ionizing Radiation, Radiotherapy, and ESR

Ionizing radiation, as the name implies, passing through a given environment causes its ionization and production of free radicals. For medical treatment, an appropriate equipment – accelerators – is used for generation such energetic irradiation. It is dangerous to living organisms and can cause mutations in DNA and its replication. Therefore, precautions must be taken, and only qualified and experienced personnel may work on the equipment that generates this radiation. To control the doses of irradiation, specific dosimeters are used.

The basic element of each organism is a cell. It is the smallest structural and functional unit of the body capable of carrying out all basic life processes. The cell is a space bounded by a cell membrane, inside which there is a number of organelles that perform various functions. One of them is a mitochondrion, where oxygen respiration occurs and reactive oxygen species (ROS) are generated.

Ionizing radiation may damage all cellular organelles; however, DNA damage is most harmful. Various changes caused in this way are possible, starting from single-strand breaks, through double-strand breaks, base damage, DNA-protein cross-links, up to multiply damaged places. All these changes can be produced by direct ionizing radiation of DNA or by the interaction of free radicals with DNA. The interaction with free radicals can be modified by free radical scavengers. There are two groups of antioxidants which are able to eliminate the harmful effect of free radicals: non-enzymatic (vitamins C, E, A, K, uric acid, coenzyme Q, selenium) and antioxidant enzymes (superoxide dismutase (SOD), catalase (CAT), glutathione peroxidase (GPx)). The most popular, present in our life scavenger of free radicals is vitamin C. Serious DNA damage causes cell death by apoptosis or necrosis. If the DNA repair systems work properly, the damage is repaired and the cell can return to its original state. The repair systems may not work properly, and then a mutation is introduced into the cell, the consequence of which may be the start of the tumor formation process (Fig. 2.2).

The type of changes caused by ionizing radiation within DNA depends on the nature of irradiation and its energy. In a case of α-particles, high ionization densities are along a linear track; for β-particles, ionization along the linear track is infrequent. Low-energy electrons make clusters of high ionization density along an irregular way. Double-strand breaks caused by high ionization are less reparable. The examples of ionizations per cell are shown in Fig. 2.3. Additionally in Table 2.1, DNA damage caused in one cell by 1Gy of X-rays is collected.

With the increasing dose of irradiation, the number of killed cells also increases. Above a certain threshold, the severity of irradiation effects is steadily increasing until the tissue or organism is completely destroyed. There is so-called threshold

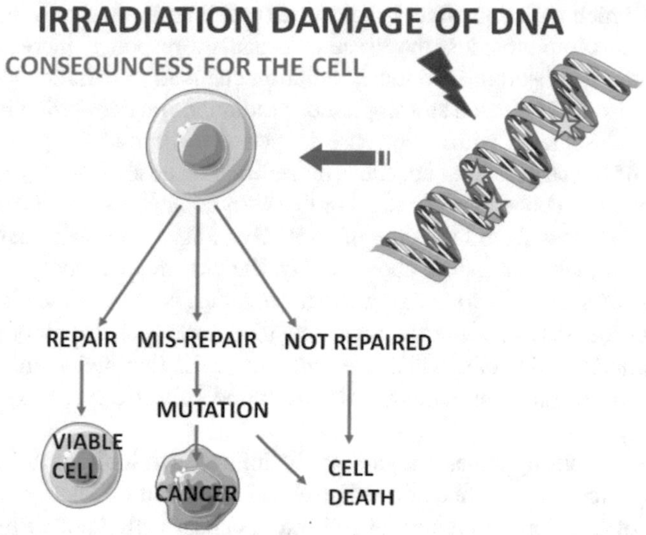

IRRADIATION DAMAGE OF DNA

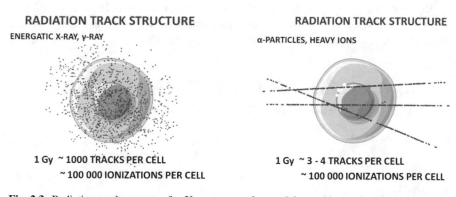

Fig. 2.2 A scheme of ionizing irradiation impact on a cell. (This figure was created using Servier Medical Art templates, which are licensed under a Creative Commons Attribution 3.0 Unported License; https://smart.servier.com)

RADIATION TRACK STRUCTURE
ENERGATIC X-RAY, γ-RAY

RADIATION TRACK STRUCTURE
α-PARTICLES, HEAVY IONS

1 Gy ~ 1000 TRACKS PER CELL
~ 100 000 IONIZATIONS PER CELL

1 Gy ~ 3 - 4 TRACKS PER CELL
~ 100 000 IONIZATIONS PER CELL

Fig. 2.3 Radiation track structure for X-, γ-rays, and α-particles and heavy ions in a cell. (This figure was created using Servier Medical Art templates, which are licensed under a Creative Commons Attribution 3.0 Unported License; https://smart.servier.com)

Table 2.1 DNA damage caused by 1 Gy of X-rays in a cell

Damage	No. per cell
Base damage	1000–2000
Cross-links	200–400
Single-strand breaks	~ 1000
Double-strand breaks	~ 40

dose below which radiation effect is balanced by cell renewal. Above this dose, the severity of the symptoms, i.e., the tissue or organism response, increases with the dose of radiation. Absorbing too much radiation can lead to radiation sickness. For example, a dose of 2 Gy, used as a single-dose radiation fraction during radiotherapy treatment, is associated with a decrease in peripheral blood lymphocyte counts. Absorption of such a dose does not cause mortality. The dose of 2–4 Gy is associated with damage to the bone marrow and 4–8 Gy with damage to the epithelium of the cellular duct. A dose from the range of 8–50 Gy is associated with damage to the nervous system, while at a dose above 50 Gy, the enzymatic activity of proteins is blocked. Tissue sensitivity to ionizing radiation varies. Nervous tissue is considered to be the most resistant, while the most sensitive are, located mainly in the bone marrow, hematopoietic cells, which are immature cells that divide and hence their high sensitivity to radiation damage. Mature blood cells are relatively resistant to radiation.

There are various hypotheses regarding the influence of ionizing radiation on the human body. According to one of them, each act of ionization leads to an increased probability of transforming a normal cell into a cancer cell. Another hypothesis is that small doses of ionizing radiation do not lead to harmful effects and even have a positive effect on the human body. And so low doses of radiation stimulate DNA repair and the immune system at the cellular level and also cause scavenging of free radicals. These processes are intended to reduce the risk of mutation or cancer.

Because our body consists mainly of water, by its radiolysis free radicals are formed (Fig. 2.4). The most popular, associated with living organisms, are reactive oxygen species (ROS), and among them are hydroxyl radical ($^\bullet$OH),

WATER IONIZATION BY GAMMA IRRADIATION
(INITIAL REACTION WITH CELLS)

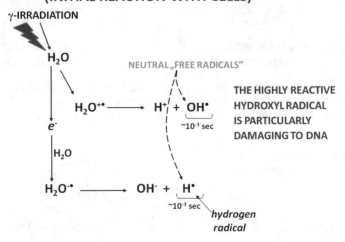

Fig. 2.4 The scheme of water ionization by γ-irradiation

superoxide radical ($O_2^{-\bullet}$), and nitric oxide ($NO^\bullet$). The first step of radiation action on water causes its ionization and then the excitation of water molecules. The excited molecules break down into hydrogen atoms and hydroxyl radicals (Fig. 2.5). Because of their high reactivity, they are able to react with other molecules which results in the multiplication of radicals or the generation of new types. Free radicals play the important role in our body, for example, they are involved in cells metabolism. In low concentration they are not harmful to the body. However, they are also important in some diseases, such as inflammatory states, cancer, or arteriosclerosis diseases. Reactive oxygen species are also involved in aging processes.

As it was mentioned previously, these molecules (paramagnetic species, including free radicals) can be measured using ESR. They are very reactive, but their lifetime is very short, and it is necessary to apply spin traps. They are substances that prolong the lifetime of free radicals, making them stable and thus detectable. There are several kinds of spin traps, and their choice depends on the type of free radicals which we want to trap. One of them is DMPO (5,5-dimethyl-1-pyrroline N-oxide) applied in the detection of hydroxyl radicals. The popular spin trap is also alpha-phenyl N-tertiary-butyl nitrone (PBN). Another method of detecting unstable radicals is freezing them and measuring at low temperatures. These solutions are mainly applied in study with blood and other tissues samples, so-called biologically active materials.

The impact of ionizing radiation on human body may be also measured in bones and tooth, due to their hydroxyapatite content. Carbonates contained in hydroxyapatite under the influence of ionizing radiation are transformed into carbonate anion radicals ($CO_2^{\bullet-}$), whose life is estimated at 10^7 years. The measurements are usually performed for powder samples, and the radiation-absorbed dose is determined by

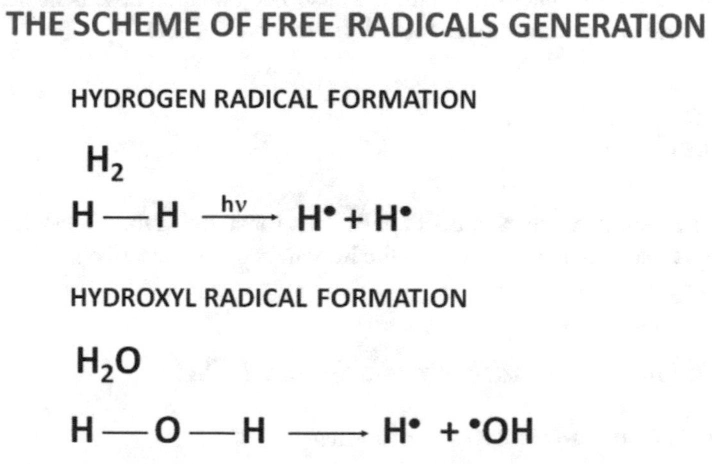

THE SCHEME OF FREE RADICALS GENERATION

HYDROGEN RADICAL FORMATION

H_2

$$H\!-\!H \xrightarrow{h\nu} H^\bullet + H^\bullet$$

HYDROXYL RADICAL FORMATION

H_2O

$$H\!-\!O\!-\!H \longrightarrow H^\bullet + {}^\bullet OH$$

Fig. 2.5 The scheme of free radicals creation

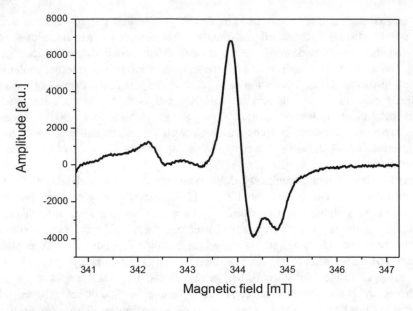

Fig. 2.6 ESR spectrum of gamma-irradiated bone

the intensity of the characteristic signal peak. The sensitivity of this measurement is estimated at 0.1 Gy.

To determine the dose absorbed during irradiation, the entire sample in the form of a powder is divided into several equal parts. Each of them is then irradiated with a specific (each one with different) dose of ionizing radiation. The concentration of generated radicals is determined for the registered ESR spectrum (Fig. 2.6). Next, a plot of the ESR signal intensity versus the absorbed radiation dose is made, and the absorbed dose is determined from the extrapolation (Fig. 2.7).

Examples

In this part, chosen examples of ESR application for studying the effects of radiation sterilization and ionizing radiation on the human body are described.

Sterilization with Ionizing Irradiation and ESR

Example 1: Gamma-Irradiated Antibiotics

The example of gamma irradiation for drugs sterilization was described by Wilczyński et al. in the paper published in 2008. The ESR spectra were mainly

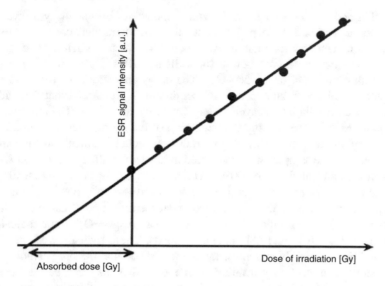

Fig. 2.7 The scheme for determining the dose absorbed during radiotherapy

recorded at room temperature in air condition. All samples were irradiated with a dose of 25 kGy. The obtained results showed the generation of free radicals in all samples; however, the differences in their concentration were observed depending on the drug substance. For penicillin derivatives, the lowest concentrations of free radicals were recorded and for aminoglycoside antibiotics – the least resistant to gamma radiation – the highest concentrations of free radicals among the samples studied. It was observed that the concentration of free radicals generated by radiation sterilization decreases with time of drugs storage.

Example 2: Steroids Gamma-Irradiated

Another example of drugs sterilization with gamma irradiation can be steroids. For irradiation, the dose of 25 kGy was applied, and the concentration of free radicals was monitored for 6 months. It was observed that free radicals were relatively durable because even after a half of year, the ESR signals were still present.

Example 3: Gamma-Irradiated Sulfonamides

ESR studies of gamma-irradiated drugs (including sulfonamides) were reported in the book chapter by Şeyda Çolak. In that report, the group of drugs that were investigated includes sulfanilamide, sulfafurazole, sulfathiazole, sulfacetamide sodium, sulfamethazine, butylated hydroxyanisole, and Albendazole. Sulfonamides are antibacterial agents used in infections of urinary system, in bacterial inflammations of the skin and eye, as well as in veterinary medicine. Butylated hydroxyanisole is used as an antioxidant in pharmaceutical preparations and cosmetic fats and oils and as a stabilizer for vitamin A. Albendazole is an anthelmintic drug. The samples were measured both before (no ESR signal) and after irradiation (ESR peaks) with different doses (5–50 kGy). The decay characteristics for these peaks were determined at normal and stability conditions. It was found that free radicals generated after irradiation decayed faster at stability conditions. The amplitudes of ESR lines fast decreased during the beginning of the storage period, and after the first days of storage, the decay rate of radiation-induced radicals in the samples was decreased. The samples were measured at low temperatures (up to 100 K), and they were also annealed (up to 400 K). The higher decay rates of the radicals were observed at high temperature than the decay rates at low temperatures. In addition to the dynamics of free radicals generated under the influence of gamma radiation, their types were also specified: SO_2^- ionic radical, $[O=S=O]^*$ ionic radical, and free radicals from S-N and C-N chemical bonds. These radicals were generated depending on material. In summary, the authors stated that the drugs studied were not sensitive to high-energy irradiation. It means that they can be sterilized by gamma irradiation with a dose up to 50 kGy because it does not cause much molecular damages. The most important conclusion is that ESR is a method for monitoring the sterilization using irradiation of drugs which contain described in this article materials as active ingredients.

Example 4: Sterilization of Chloramphenicol Antibiotic

The effect of ionizing irradiation on physical and chemical properties of chloramphenicol in solid state were studied using various methods scanning electron microscope (SEM), X-ray, chromatography (TLC), UV (ultraviolet) and IR (infrared) spectrophotometry, thermal (differential scanning calorimetry, DSC), and of course electron spin resonance (ESR). Chloramphenicol is an antibiotic characterized by a wide spectrum antibacterial activity. The doses of applied irradiation were in the range from 25 to 400 kGy. From ESR measurements, the signals from free radicals generated by the irradiation were analyzed. The increase in signal intensity with increasing dose of irradiation was observed, and the free radicals were stable longer than 6 months. Although ionizing radiation generates free radicals, no significant changes in chloramphenicol properties were observed.

Example 5: Disulfiram, Radiosterilization, and ESR

Another example of radiosterilization was described by Marciniec et al. in the paper from 2011. In this study, disulfiram (drug applied in treatment of alcoholism to prevent unpleasant side effects, e.g., headaches) was irradiated with the doses of 10–100 kGy, and its physicochemical properties were studied using various methods, including ESR. Free radicals generated by ionizing radiation disappeared after about a year. The authors stated that disulfiram can be subjected to sterilization by irradiation with maintaining its physicochemical properties.

Radiotherapy, Ionizing Irradiation, and ESR

Example 6: Reconstructions of Radiotherapy Doses from Tooth Enamel Using ESR

Electron spin resonance was used as the retrospective dosimetry of doses absorbed in patients undergoing radiotherapy. ESR signals were recorded in teeth extracted from six patients within a few years after radiotherapy treatment with electron beams, ^{60}Co photons, and high-energy photon. The dose reconstructions from ESR measurements were determined with the accuracy of 5–9%. The differences between doses calculated in radiotherapy treatment planning and these obtained from ESR measurements were a few percent for teeth located within the irradiated field up to 120% for teeth placed outside the beam. It can be explained by changes in geometry of oral tissues.

Example 7: Chernobyl Accident – ESR as a Dosimeter

ESR method was also used after Chernobyl accident. The effects of ionization were measured in tooth enamel samples. The advantage of such study is undoubtedly that it can be used at any time after the exposure for dose reconstruction and the low limit of sensitivity of 0.1 Gy. It was also shown that it is better to use tooth enamel samples than blood samples, which are applicable only in a short time after the exposure. Samples were properly prepared for ESR measurements, taking into account chemical treatment, removal of the remainder of the dentine, crushing, and purification. The authors mentioned about some problems connected with ESR dosimetry, such as the effect of enhanced sensitivity to low-energy photons (overestimation of the tissue-absorbed dose) and generation of paramagnetic centers by UV light. Summarizing the results described in these articles, the authors stated that ESR dosimetry needs further investigation.

Example 8: Tooth Enamel as ESR Dosimeter

The application of tooth enamel as a detector material in dosimetry using ESR was also described by Egersdörfer et al. in 1996. In such studies, the problem is a broad background signal which can be reduced by high microwave power (saturation). The samples for the determination of irradiation dose should be carefully prepared because any action on the sample during processing may distort the final result.

The fitting procedure for irradiation dose is also important, and it may affect the correctness of the final result.

Example 9: Breast Cancer, Radiotherapy Treatment, and ESR

Ceruloplasmin, which contains Cu^{2+} ions, and transferrin, which contains Fe^{3+} ions, can be studied using ESR, because these molecules are proteins, which play a potential role in breast cancer. In this study, the authors set themselves the goal of checking how the paramagnetic centers in the blood of patients with breast cancer change under the influence of radiation therapy. ESR was also applied to study the impact of radiotherapy treatment in women with breast cancer. In that study, the samples of peripheral blood were taken before and after irradiation with a dose of 2.5 Gy and ESR measured. Based on the analyzed data, differences in ESR spectra were found recorded before and after radiotherapy. The intensity of the ESR signal from Cu^{2+} in ceruloplasmin significantly decreased in all patients after the delivery of the radiation fraction. The changes in intensity of the signal from Fe^{3+} in transferrin were divided in three groups: a significant increase, a significant decrease, or an insignificant change after the irradiation. The authors stated that such changes in ESR signal after radiotherapy treatment may indicate how the organism adapts to the biochemical stress related to the disease and how it fights against cancer disease. The representative ESR spectra are shown in Fig. 2.8.

Example 10: TEMPO Spin Label, Radioprotector, and ESR

One of the most interesting researches described in literature is the study of spin labels application as radioprotectors in radiotherapy. Spin labels are model free radicals which characteristics are their stability and durability over time. They are ideal samples to be studied using ESR. Because ionizing irradiation generates free radicals, spin label radicals scavenge them, and this can be observed in ESR spectrum (changes in intensity of ESR signal).

Yukio Nagasaki connected chosen spin labels with triblock copolymer and developed radical containing polymer nanoparticles (micelles). Attached in this case chosen TEMPO (a spin label) is used as reactive oxygen species (ROS) scavenger.

In one of the papers, the authors applied such polymer nanoparticles as radioprotectors and administered them to mice before irradiation. They observed that nanoparticles have indeed protected the animals against radiation-induced mortality. Such nanoparticles are able to target ROS and, thanks to their structure and properties, to retain longer in the body and thus to prolong their radioprotective efficacy.

In another paper by Yukio Nagasaki, the authors used such redox nanoparticles to reduce organ dysfunctions and death in whole-body-irradiated mice and applied ESR method to control the stability and persistence in the blood circulation of such

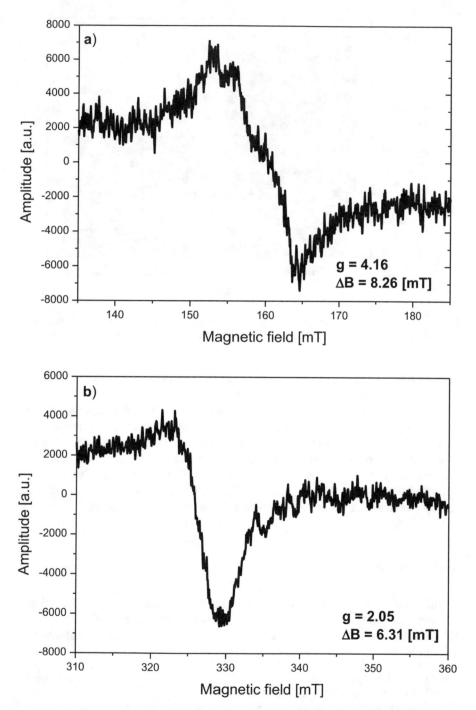

Fig. 2.8 Representative ESR spectra of whole blood from person with breast cancer treated with radiotherapy: (**a**) from Fe^{3+} and (**b**) from Cu^{2+} ions

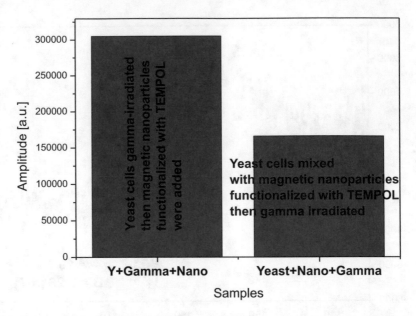

Fig. 2.9 Concentration of free radicals in gamma-irradiated and incubated yeast cells previously without (left graph) and with (right graph) TEMPOL-functionalized magnetic nanoparticles

nanoparticles. For this purpose, blood samples with nanoparticles were studied, and the changes in ESR spectra were analyzed.

Similar results were shown during the 8th Annual World Congress of Nano Science and Technology in Potsdam, 2018. In this case, yeast cells were incubated with TEMPOL-functionalized magnetic nanoparticles and gamma-irradiated. It was observed that for yeast cells with such nanoparticles, the effect of irradiation was less harmful, and the concentration of free radicals was lower than for yeast cells irradiated without nanoparticles (Fig. 2.9). This result also confirms the radioprotective effect of nanoparticles functionalized with spin labels.

Conclusions

In this chapter, the application of ESR in healthcare and pharmaceutical science is mentioned. In these fields, ESR can be applied to study free radicals generated by ionizing irradiation used for sterilization and generated during radiotherapy treatment.

Study Questions
1. What methods of sterilization do you know?
2. For which materials ionizing radiation sterilization is used?

3. What is the advantage of ionizing radiation sterilization compared to gaseous methods?
4. How the effect of ionizing radiation sterilization is checked by electron spin resonance?
5. What is meant by "ionizing irradiation"?
6. What kind of free radicals can be created by ionizing irradiation?
7. What is the difference between spin label and spin trap? How they can be applied in the study with electron spin resonance?

References

Y. Sultana, Pharmaceutical Microbiology and Biotechnology, Sterilization Methods and Principles, New Delhi, 2007

S. Wilczyński, B. Pilawa, M. Ptaszkiewicz, J. Swakoń, P. Olko, Free radicals properties of gamma irradiated solid forms of drugs, Engineering of Biomaterials, 81–84 (2008) 52–54

M. Gibella, A-S. Crucq, B. Tilquin, P. Stocker, G. Lesgards, J. Raffi, Electron spin resonance of some irradiated pharmaceuticals, Radiation Physics and Chemistry 58 (2000) 69–76

J.P. Basly, I. Basly, M. Bernard, Electron spin resonance identification of irradiated ascorbic acid: Dosimetry and influence of powder fineness, Analytica Chimica Acta 372 (1998) 373–378.

G.P. Jacobs, P.A. Wills, Recent developments in the radiation sterilization of pharmaceuticals, Radiation Physics and Chemistry 31(1988) 685–691.

A. Engalytcheff, R. Debuyst, G.C.A.M. Vanhaelewyn, F.J. Callens, B. Tilquin, Attempts at Correlation of the Radiolytic Species of Irradiated Solid-State Captopril Studied by Multi-frequency EPR and HPLC, Radiation Research 162 (2004) 616.

H. Terryn, V. Deridder, C. Sicard-Roselli, B. Tilquin, C. Houee-Levin, Radiolysis of proteins in the solid state: an approach by EPR and product analysis, Journal of Synchrotron Radiation 12 (2005) 292.

W. Bögl, Radiation sterilization of pharmaceuticals – chemical changes and consequences, Radiation Physics and Chemistry 25 (1985) 425–435.

B. Marciniec, G. Przybytniak, M. Ogrodowczyk, EPR study of some E-beam irradiated steroids in the solid state. Annals of Polish Chemical Socety II (2005) 238–241.

I. Porębska, B. Sokołowska, S. Skąpska, S.J. Rzoska, Treatment with hydrostatic pressure and supercritical carbon dioxide to control Alicyclobacillus acidoterrestris spores in apple juice, Food Control 73 (2017) 24–30

B. Windyga, M. Rutkowska, B. Sokołowska, S. Skąpska, A. Wesołowska, M. Wilińska, M. Fonberg- Broczek, S.J. Rzoska, Inactivation of Staphylococcus aureus and native microflora of breast milk by high pressure processing. High Pressure Research 35(2) (2015) 181–188.

Amin I. Kassis, S. James Adelstein, Radiobiologic Principles in Radionuclide Therapy, The Journal of Nuclear Medicine, 46(1) (2005) 4S–12S

T. Sandle, Steam sterilisation, Sterility, sterilisation and sterility assurance Published byWoodhead Publishing Limited 2013a, 93–109.

W.J. Rogers, Steam and dry heat sterilization of biomaterials and medical devices, Sterilisation of Biomaterials and Medical Devices, Woodhead Publishing Limited 2012, 20–55.

T. Sandle, Dry heat sterilisation, Sterility, sterilisation and sterility assurance Published by Woodhead Publishing Limited 2013b, 83–92.

T. Sandle, Gaseous sterilisation, Sterility, sterilisation and sterility assurance Published by Woodhead Publishing Limited 2013c, 111–128.

G.C. Mendes, T.R.S. Brandǎo, C.L.M. Silva, Ethylene oxide (EO) sterilization of healthcare products, Sterilisation of Biomaterials and Medical Devices, Woodhead Publishing Limited 2012, 71–96.

S. Iwaguch, K. Matsumura, Y. Tokuoka, S. Wakui, N. Kawashima, Sterilization system using microwave and UV light, Colloids and Surfaces B: Biointerfaces 25 (2002) 299–304.

M. Tohfafarosh, D. Baykal, J.W. Kiel, K. Mansmann, S.M. Kurtz, Effects of gamma and e-beam sterilization on the chemical, mechanical and tribological properties of a novel hydrogel, Journal of the Mechanical Behavior of Biomedical Materials 53 (2016) 250–256.

M. Walo, G. Przybytniak, A. Nowicki, W. Świeszkowski, Radiation-Induced Effects in Gamma-Irradiated PLLA and PCL at Ambient and Dry Ice Temperatures, Journal of Applied Polymer Science Vol. 122 (2011) 375–383.

G. Przybytniak, E.M. Kornacka, K. Mirkowski, M. Walo, Z. Zimek, Functionalization of polymer surfaces by radiation-induced grafting, Nukleonika 53(3) (2008) 89–95.

R.F. Morrissey, C.M. Herring, Radiation sterilization: past, present and future, Radiation Physics and Chemistry 63 (2002) 217–221.

B. Marciniec, K. Dettlaff, Radiation sterilization of drugs. Trends in radiation sterilization, Vienna: IAEA; 2008.

J.W. Hopewell, Late radiation damage to the central nervous system: a radiobiological interpretation. Neuropathology and Applied Neurobiology 5(5) (1979) 329–343.

Z. Zdrojewicz, A. Szlagor, M. Wielogórska, D. Nowakowska, J. Nowakowski, Influence of ionizing radiation on human body, Family Medicine & Primary Care Review 18(2) (2016) 174–179

M. Valko, M. Izakovic, M. Mazur, C.J. Rhodes, J. Telser, Role of oxygen radicals in DNA damage and cancer incidence, Molecular and Cellular Biochemistry 266 (2004) 37–56.

C. Chatgilialoglu, P. O'Neill, Free radicals associated with DNA damage, Experimental Gerontology 36 (2001) 1459–1471.

S. Rana, R. Chawla, R. Kumar, S. Singh, A. Zheleva, Y. Dimitrova, V. Gadjeva, R. Arora, S. Sultana, R. Kumar Sharma, Electron paramagnetic resonance spectroscopy in radiation research: Current status and perspectives, Journal of Pharmacy and Bioallied Sciences 2(2) (2010) 80–87.

H.M. Swartz, N. Khan, J. Buckey, R. Comi, L. Gould, O. Grinberg, A. Hartford, H. Hopf, H. Hou, E. Hug, A. Iwasaki, P. Lesniewski, I. Salikhov, T. Walczak, Clinical applications of EPR: overview and perspectives, NMR in Biomedicine 17 (2004) 335–351

Edited by M. Nenoi, Evolution of Ionizing Radiation Research, 2015, IntechOpen, S. Çolak, chapter 12: Ionizing Radiation Used in Drug Sterilization, Characterization of Radical Intermediates by Electron Spin Resonance (ESR) Analyses, pp. 281–306

T. Winiecki, J. Kazmierska, R. Krzyminiewski, B. Dobosz, Z. Kruczynski, T. Kubiak, Utility of EPR for evaluation of free radicals and iron complexes in blood in patients before and after radiotherapy, Radiotherapy and Oncology Volume: 103, Supplement 1, May, 2012, pp. S574

R. Krzyminiewski, B. Dobosz, T. Kubiak, The influence of radiotherapy on ceruloplasmin and transferrin in whole blood of breast cancer patients, Radiation and Environmental Biophysics 56 (2017) 345–352

Chitho P. Feliciano, Y. Nagasaki, Antioxidant Nanomedicine Protects against Ionizing Radiation-Induced Life-Shortening in C57BL/6J Mice, ACS Biomaterials Science and Engineering 5, 11 (2019) 5631–5636

Chitho P. Feliciano, K. Tsuboi, K. Suzuki, H. Kimura, Y. Nagasaki, Long-term bioavailability of redox nanoparticles effectively reduces organ dysfunctions and death in whole-body irradiated mice, Biomaterials 129 (2017) 68–82

B. Dobosz, R. Krzyminiewski, G. Schroeder, J. Kurczewska, Magnetite Nanoparticles as Drug Carriers in Cancer Treatment, BIT'S 8th Annual World Congress of Nano Science & Technology, 24–24 October 2018, Potsdam, Germany

A.E. Baranov, A.K. Guscova, N.M. Nadejina, V.Yu. Nugis, Chernobyl Experience: Biological Indicators of Exposure to Ionizing Radiation, Stem Cells 13(suppl 1) (1995) 69–77

V.V. Chumak, I.A. Likhtarev, S.S. Sholom, L.F. Pasalskaya, Y.V. Pavienko, Retrospective Reconstruction of Radiation Doses of Chernobyl Liquidators by Electron Paramagnetic Resonance,

Armed Forces Radiobiology Research Institute Bethesda, Maryland, USA, Editor and NIS Initiatives Coordinator Glen I. Reeves, M.D., 1997

S. Egersdörfer, A. Wieser, A. Müller, Tooth Enamel as a Detector Material for Retrospective EPR Dosimetry, Applied Radiation and Isotopes 47(11/12) (1996) 1299–1303

B. Marciniec, M. Stawny, M. Kozak, M. Naskrent, The effect of ionizing radiation on chloramphenicol, Journal of Thermal Analysis and Calorimetry, Vol. 84 (2006) 3, 741–746

C.A. Rice-Evans, A.T. Diplock, M.C.R. Symons, Laboratory techniques in biochemistry and molecular biology – techniques in free radical research 1977, pp. 51–99

S.A. Dikanov, A.R. Crofts, Electron paramagnetic spectroscopy. Handbook Applied Solid State Spectroscopy 2006, 97–149

D. Freude, Chapter 3 – Paramagnetic electron resonance, Spectroscopy 2006, June 1–14

J.A. Weil, J.R. Bolton, Electron Paramagnetic Resonance: elementary theory and practical applications, second ed., Wiley Interscience, Hoboken, 2007

H.M. Swartz, G. Burke, M. Coey, E. Demidenko, R. Dong, O. Grinberg, J. Hilton, A. Iwasaki, P. Lesniewski, M. Kmiec, Kai-Ming Lo, R. Javier Nicolalde, A. Ruuge, Y. Sakata, A. Sucheta, T. Walczak, B.B. Williams, Ch.A. Mitchell, A. Romanyukha, D.A. Schauer, In vivo EPR for dosimetry, Radiation Measurements 42 (2007) 1075–1084

B. Ciesielski, A. Karaszewska, M. Penkowski, K. Schultka, M. Junczewska, R. Nowak, Reconstruction of doses absorbed by radiotherapy patients by means of EPR dosimetry in tooth enamel, Radiation Measurements 42 (2007) 1021–1024

D. Adjei, A. Wiechec, P. Wachulak, M.G. Ayele, J. Lekki, W.M. Kwiatek, A. Bartnik, M. Davídková, L. Vyšín, L. Juha, L. Pina, H. Fiedorowicz, DNA strand breaks induced by soft X-ray pulses from a compact laser plasma source, Radiation Physics and Chemistry 120 (2016) 17–25

S. Suzen, H. Gurer-Orhan, L. Saso, Detection of Reactive and Nitrogen Species by Electron Paramagnetic Resonance (EPR) Technique, Molecules 22 (2017) 181

I.U. Ahad, B. Butruk, M. Ayele, B. Budner, A. Bartnik, H. Fiedorowicz, T. Ciach, D. Brabazon, Extreme ultraviolet (EUV) surface modification of polytetrafluoroethylene (PTFE) for control of biocompatibility, Nuclear Instruments and Methods in Physics Research B 364 (2015) 98–107

A.K.Shukla (Ed.) Electron Magnetic Resonance – Applications in Physical Sciences and Biology. Experimental Methods in the Physical Sciences, Volume 50, Academic Press (an imprint of Elsevier), 2019, B. Dobosz, R. Krzyminiewski, EMR in geology/mineralogy, pp. 21–39, ISBN: 978-0-12-814024-6, ISSN: 1079-4042

B. Marciniec, K. Dettlaff, M. Naskrent, Z. Pietralik, M. Kozak, DSC and spectroscopic studies of disulfiram radiostability in the solid state, Journal of Thermal Analysis and Calorimetry (2011) DOI https://doi.org/10.1007/s10973-011-1810-4

Chapter 3
ESR Applications in Paleontology and Geochronology

Natural Radioactivity and the Accumulation of Radiation Radicals

Soil represents an almost ubiquitous formation consisting of mineral and organic components, the first one being generated during entire Earth history by alteration and recirculation mainly of the superior solid crust. Soil appears as an unconsolidated medium in a continuous interaction with atmosphere, hydrosphere as well as biosphere and offers shelter for an immense diversity of living organism and, in a certain measure, a medium for preservation and long-time conservation of their remnants as fossil.

As the Earth crust represents the main mineral source of the soil, in its composition entered the same radioactive elements that now can be found in the Earth Upper Continental Crust and of which lifetime is comparable with the Earth age, now estimated at 4.54 ± 0.05 billion years.

Accordingly, the main radioactive elements which contribute to soil radioactivity in a proportion greater than 99% are ^{40}K, ^{232}Th and ^{238}U and at a lesser extent ^{235}U. It is worth mentioning that ^{232}Th, ^{235}U and ^{238}U represent star member of homonyms radioactive chain decay series which maintain the presence of another 10, 14 and, respectively, 16 other radioactive elements, all of them emitter of high ionizing α, β and γ rays able to induce a multitude of radioactive damages in the form of free radicals sometimes called radiation centres in neighbouring materials. In solids, radiation-induced free radicals usually consist of a pair of a hole and an electron in the form of an interstitial atom and a vacancy which are trapped into crystalline lattice or at the place of an impurity (a foreign ion, different from those which form the main mineral).

With contributions by Octavian G. Duliu, Vasile Bercu.

Depending on chemical and mineralogical composition of mineral constituents of the soil, the lime-life of such free radicals can be up to hundreds of thousands of years or even of millions of years making their concentration potential age proxies. This fact is especially useful in the case of Quaternary (from 2.588 10^6 years to present) geochronology where the classical methods, based on the decay of long-lived radionuclides such as ^{40} K (potassium-argon and argon-argon geochronology) or ^{232}Th and ^{235,238}U (thorium-lead and uranium-lead geochronology), are not suitable due to shortest ages with respect to considered radionuclides time-life (1.251, 14,05, 0.738 and 4.468 10^9 years, respectively).

For these reasons, excepting dendrochronology, with an age record of 19 ky, and ^{14}C, with an age limit of about 30 ky, the only suitable methods are those based on the uranium isotopes' disequilibrium and especially those based on the accumulation of radiation free radicals or radiation centres (Fig. 3.1). As a general feature, regardless of their structure, all are generated by the action of ionization radiation emitted by the radioactive elements inside and outside the sample as in the case of fossils or only by the radioactive elements contained by the sample as in the case of speleothems, and the radiation-induced free radicals continuously accumulate. In this case, the age of the sample could be determined by comparing the amount of free radicals accumulated, since fossil buried in soil or speleothem formation, with the annual rate of the free radicals' generation. The age of the sample can be evaluated by dividing the concentration of existing free radicals to the annual ratio of their generation (see next paragraphs).

In this regard, it is worth mentioning that the content of radiation free radicals is proportional to the amount of energy absorbed from the radiation emitted by natural radionuclides, the so-called paleodose (PD). The content of radioactive elements is well determined by different spectroscopic methods such as X-ray fluorescence, high-resolution gamma ray spectrometry, instrumental neutron activation analysis or inductively coupled mass spectrometry; therefore, the amount of energy annually absorbed by the material, the so-called annual dose (AD), is at present time well established and published in different papers or textbooks as content to dose-rate conversion factors (Table 3.1).

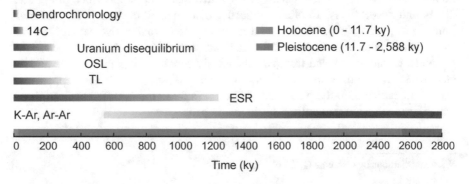

Fig. 3.1 The limits of Quaternary absolute geochronology methods

Table 3.1 The annual dose (in 10^{-3} Gy/y) due to the natural content of one mg/kg of ^{232}Th, ^{235}U $+^{238}$U and 1% of potassium

Element	Annual dose (10^{-3} Gy/y)	Remarks
^{40}K	1.0473	
^{232}Th	0.8131	Entire series
^{235}U $+^{238}$U	3.0523	Both U series

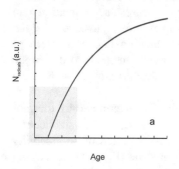
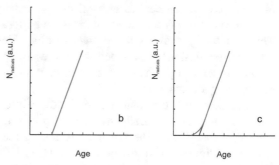

Fig. 3.2 Time evolution of the radiation free radicals' concentrations in the case of long ages or great dose rates (**a**). The inset reproduced at magnified scale illustrates for small ages or reduced dose rates a pure linear dependency (**b**) or the effect of supra-linear dependency (**c**), in which case, the extrapolation of linear dependency gives smaller ages

Although remarkably stable, the radiation free radicals have a finite time-life so that the kinetic of their time evolution can be better described by a first-order differential equation as follows:

$$\frac{dN}{dt} = R - ln2\frac{t}{T_{1/2}} \tag{3.1}$$

where N represents the content of radiation free radicals, R represents the annual rate of their generation and $T_{1/2}$ is the corresponding half-life time.

The general solution of Eq. (3.1) represents a saturation exponential illustrated in Fig. 3.2a:

$$N(t) = R\frac{T_{1/2}}{ln2}\left[1 - \exp\left(-ln2\frac{t}{T_{1/2}}\right)\right] \tag{3.2}$$

If the sample age is significantly smaller than the irradiation free radicals' half-life time, Eq. (3.2) becomes:

$$N(t) = Rt \tag{3.3}$$

The radiation free radicals' kinematics described by Eq. (3.2) could be regarded as an ideal situation which applies in the case of high dose rated and great ages. In the case of low ages, sometime the radiation radicals' content vs. dose curve differs from a linear dependency, a behaviour known as supra-linearity which can give a lower age than the real one (Fig. 3.2b, c).

It can be remarked that both solutions of Eq. (3.1) show that the irradiation defects monotonously accumulate during the sample history, provided that the content of radioactive elements around and within the sample remained unchanged. By taking into account that the half-life time of main natural radionuclides ^{40}K, ^{232}Th and ^{235,238}U are of the order of 10^9 years with respect to sample ages which are of the order of maximum 10^6 years, this supposition is perfect valid. This fact makes possible to calculate the sample aged by determining the actual content of radiation centres as well as the radiation centres' accumulation rate given the content of radioactive elements in both samples and neighbouring media. This peculiarity converts the problem of age determination into a problem of retrospective dosimetry, provided that there are appropriate methods to determine with accuracy and precision the content of radiation radicals/centres as well as the rate of their accumulation.

At present time, the content of radiation centres can be determined by thermo-luminescence (TL), optically stimulated luminescence (OSL) as well as electron spin resonance (ESR), the last one presenting the longest period of applicability reaching a time span from Holocene to Middle Pleistocene (Fig. 3.1). Regarding EPR absolute geochronology, it is worth mentioning that the optimum ages are between 30 and 250 ky, greater than the maximum ages which could be confidently determined by ^{14}C and minimum ages for which K-Ar and Ar-Ar give positive results. Only in some instances, the uranium isotopes' disequilibrium could be used for ages no greater than 200 ky.

It should be pointed out that the age determined by one of these methods should be understood as absolute age from the moment the geochronological clock was reseated which could be the moment of fossil burial or the moment when the rock temperature was low enough to stop radiation defects' recombination or, as the case of speleothem, the calcite recrystallizes.

On the other hand, K-Ar and Ar-Ar are used to date only volcanogenic materials such as lava, tephra or pumice, while uranium series disequilibrium gave good results in the case of some speleothems and marine sedimentary samples such as manganese nodules. On contrary, the geochronology based on the accumulation of irradiation centres showed to be the most suitable. In this regard, ESR showed a large domain of materials including mammal fossil teeth and other bones, fragments of mollusc shell, speleothems, carbonate concretions (caliches) corals, foraminifera, volcanic quartz bearing rocks or optically bleached quartz containing materials.

As the irradiation free radicals are generated by the ionizing high energy radiation emitted by the radioactive elements existing within the sample as well as in its immediate vicinity, the most important problem consists in an accurate estimation of the contribution of different radiation to the radiation radicals' accumulation.

As mentioned before, all natural radioactive elements as well as their daughter emit during and post disintegration corpuscular alpha and beta as well as electro-magnetic gamma radiation.

Alpha radiation are emitted by the member of natural radioactive series and have energies around 5 MeV. Beta radiation is also emitted by the member of radioactive series as well as by the ^{40}K with energies up to 1 MeV. For the same radioactive elements, including ^{40}K, the maximum energy is about 2.6 MeV. Consequently, depending on the radiation mass, charge and energy, their maximum linear path into material will vary from tens of microns for alpha ray to few of millimetres in the case of beta radiation and tens of centimetres for gamma rays.

In this regard, it should be evidenced that both alpha and beta radiation are radiation with direct ionizing and consequently have a limited path into material, while the gamma ray which is a radiation with indirect ionization presents an exponential attenuation which practically means an infinite path. This means that the radiation free radicals are generated by both alpha and beta rays in closer vicinity of emitting radionuclide, while gamma ray can induce radiation free radicals within a large radius of tens of centimetres around radioactive nucleus.

For this reason, in the case of gamma rays, the maximum linear path defined for alpha and beta radiation as the maximum thicknesses of a materiel necessary to attenuate completely these radiations is replaced by the mean path, i.e. the material thickness which reduces the flux of gamma ray by e times, where $e =$ 2.718...represents the basis of natural logarithms.

All these particularities should be taken into account when the size of the examined sample is of the order of magnitude of alpha, beta or gamma rays emitted by the neighbouring radioactive elements.

ESR Age Determination

ESR geochronology can be classified as a particular case of retrospective dosimetry. The main idea behind these techniques consists of investigating the accumulation of different long-lived free radicals in a great variety of solid materials under the influence of ionizing radiation emitted by natural radioactive elements within the sample and neighbouring media.

For this reason, the age of a sample can be calculated by comparing the concentration of radiation centres in investigated material with the annual dose rate due to the radioactive elements as mentioned before. To achieve this goal, one can apply different methods; however, among the most widespread and with the best results, we mention here the additive dose method and the reconstructive dose techniques, which are very similar to those used by thermoluminescence and optically stimulated luminescence geochronology.

Reconstructive Dose Method

Equation (3.2) and Fig. 3.2 provide a typical case of radiation defects' accumulation when the saturation appears only due to recombination processes and the defect concentration-dose curve is well described by the a saturation exponential function. In this case, by considering the age t_0 of the sample, the concentration n_0 of the radicals induced by ionizing radiation, which is found in present, will correspond to a total absorbed dose of the sample since from the beginning until today, i.e.

the paleodose *PD* mentioned before is still unknown. As its value contains entire information of sample age, the main problem for the ESR geochronology consists in its accurate determination.

Accordingly, in view of a presumed proportionality of the absorbed dose rate with the exposure time, Eq. (3.2) can be rewritten by replacing the exposure time with the absorbed dose. In this way, the curve reproduced in Fig. 3.2 will describe the dependency of the concentration of radiation radicals by the absorbed dose, which, in the hypothesis of a constant of ambient radioactive elements, the annual dose debt will be monotonously increased with the exposure time (see Fig. 3.3).

This fact raises the problem of the maximum ages which could be determined by ESR geochronology. Given that the content of radiation free radicals follows a saturation exponential dependency, it is reasonable to consider the maximum age as equal to 0.2–0.25 of the radiation centres' lifetime.

Additive Dose Method In these conditions, one of the best methods to determine the age of a sample consists in using the additive dose method. Within this method, after determining the intensity of ESR spectrum in the present, the sample is progressively irradiated with well-known doses. After each irradiation, the resulting ESR spectrum is recorded to determine the ESR intensity given by the radiation radical chosen for analysis. The resulting set of data, i.e. the growth function, is fitted with a linear or a saturation exponential function.

The absolute age of the sample is determined, as mentioned before, by dividing the resulted *PD* to the annual dose rate as it results from the content of natural radioactive radicals which contribution to annual dose is those mentioned in Table 3.1. This procedure is illustrated in Fig. 3.3.

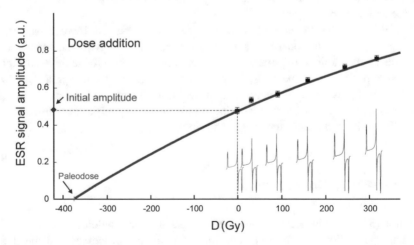

Fig. 3.3 The additive dose method to calculate the PD. The initial ESR signal (red ink) is recorded followed by the irradiation of the same sample at increased doses. The experimental amplitude dose growth curve is then fitted with a linear or a saturation exponential curve. The PD is equal to the intercept of the growth curve

Depending on the accuracy of ESR determinations, the apparent relative errors of the ESR age are around 0.1%, but this uncertainty regards only the fitting of saturation curve considering that the additive doses are determined with an inaccuracy of about 0.3%. In these conditions, the upper theoretical age of the ESR geochronology is limited by the signal long-term stability as well as by the saturation effects of the radiation centres in the material. As the saturation dose is about 10 kGy, given the annual dose debit of about 20 mG/y, it results in a maximum theoretical age of 0.5 My.

This maximum age should be regarded with precaution as it is due to the total energy of alpha, beta and gamma rays deposited in the material in the conditions that the maximum linear path of radiation varies, as mentioned before, between few microns and few millimetres for alpha and, respectively, beta rays, while in the case of gamma ray, the mean path is of the order of tens of centimetres. Also the size as well as the distribution of radionuclides around and within the sample could significantly influence the maximum age which could be determined by ESR geochronology.

In practice, the average error in age calculation is of 15 to 25%, relatively high for the absolute geochronometry where errors in some cases are lower than 0.5%. This error was estimated by comparing ESR ages with the ages obtained by other absolute geochronological methods. The main explication of this peculiarity is related to the unknown behaviour of the radiation centres for greatest age as well as on the possible long-term variation of the content of natural radionuclides in soil or other conservation media.

These uncertainties are more important in the case of open systems such as tooth enamels fossilized in soil where the uranium content, due to hexavalent uranium solubility, can fluctuate in large limits which significantly alters the ESR signal vs. dose curve. For this reason, it is recommendatory to use whenever it is possible other alternative geochronological methods.

Regenerative Dose Method In the regenerative dose method, the PD is not obtained anymore by fitting the ESR signal intensity generated by laboratory dose but by reconstructing the entire growth curve of the ESR signal. As in the case of additive method, in the case of regenerative one, the ESR signal is recorded before any supplementary irradiation. Then, different from the additive method, the sample is heated until no ESR signal is detected (usually at temperature higher than 400°C for more than 30 min). As a result, all free radicals able to generate an ESR signal are recombined. After thermal annealing, the same sample is irradiated in the laboratory condition with progressively increasing dose, the ESR signal being recorded in the same conditions until the ESR signal intensity overpass the initial, un-annealed one.

The resulting series of signal intensities are interpolated, and the value of *PD* is simply obtained by identifying the laboratory dose which corresponds to the intensity of the ESR signal found in the sample before annealing. As in the previous case, the essence of this method is illustrated in Fig. 3.4.

The new approach in the ESR dating represents an adaptation of a previously developed technique currently used in optically stimulated luminescence (OSL)

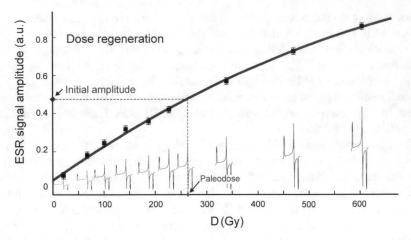

Fig. 3.4 The regenerative dose method to calculate the PD. The initial ESR signal (red ink) is recorded and reset by high-temperature annealing. After that, the same sample is irradiated at increased doses followed by reconstruction of the characteristic growth curve. The PD is determined by intersecting the reconstructed growth curve with the horizontal line corresponding to the initial ESR signal amplitude

absolute geochronology, especially in the case of single aliquot regenerative dose (SAR) protocol.

Applied to ESR absolute geochronology, this method presents a series of advantages with respect to the additive dose method:

(i)   the PD value is obtained by interpolation and not by fitting the ESR signal, i.e. the PD value is not depending any more on the fitting function
(ii) by using single aliquot experiment, one can avoid the inter-aliquot scattering in the additional irradiation process
(iii) using different steps in protocols, e.g. preheat plateau, one can increase the saturation dose and in consequence can date older samples
(iv) by using a single aliquot, the age of the same sample can be calculated as an average of multiple determination performed for different aliquots which significantly increases the accuracy.

In this regard, it is worth mentioning that, with respect to OSL dating, in the case of the ESR method, the saturation appears at much higher doses allowing significant increase of PD to be determined which increases the ESR dating age interval to an upper limit of 2.5 Ma, i.e. the Gelasian (1.8–2.58 Ma), the inferior stage of Pleistocene.

Media Frequently Dated by ESR

As mentioned before, the ESR geochronology is currently used to determine the absolute age of a multitude of samples, mainly consisting of calcium carbonate in the case of speleothem, carbonaceous concretion in loess, coral or mollusc fossils, phosphates as bone or tooth enamel hydroxyapatite as well as quartz extracted from natural rocks such as granite or some sedimentary rocks, the loess being one of the most representative media.

Carbonates Natural (bioorganic and sedimentary) carbonates are one of the most common systems of which age can be determined by ESR spectroscopy. Natural carbonates mainly form the polymorph encountered: mineral aragonite and mineral calcite. The third calcium carbonate polymorph, the vaterite, is very rare and, as the aragonite, slowly turns into calcite.

Dolomite a natural carbonate of Mg and Ca with the general formula $MgCa(CO_3)_2$ can be regarded as a solid mixture of magnesium and calcium carbonates so that the structure of radiation radicals is more complicated which makes dolomite less used in ESR geochronology. Moreover, as the ionic radii of Mg^{2+} and Mn^{2+} are very close, sometime the dolomite has a such high quantity of manganese which makes the dolomite again unsuitable for ESR geochronology.

Natural calcium carbonates can be of bioorganic nature such as coral exoskeleton of bivalve or gastropods' external shells, formed by recrystallization from aqueous solution as the speleothems or calcretes or hydrothermal in the case of travertine. In this regard, it is worth mentioning that both bioorganic calcium carbonate and speleothems are formed by slow recrystallization making which make them very pure and, thus, very convenient for ESR geochronology.

Although, in calcium carbonates, the high-energy ionizing radiation such as those emitted by natural radionuclides can generate more free radicals, e.g. CO_2^-, CO_3^- and CO_3^{3-}, as well as impurity-associated SO_3^- and SO_2^-. The CO_2^- rapid rotating radiation radical is characterized by a single, almost symmetric resonance ESR, line having the g-factor equal to 2.0006. All these centres can be used to determine the absolute age of the specimen. As a rule, natural carbonates display a relatively intricate ESR spectrum consisting of more lines associated with the above mentioned free radicals among which, the CO_2^- centre characterized by an almost isotropic resonance line with g-factor equal to 2.006 gave one of the best results.

A problem which was encountered in the case of speleothems is connected with the disequilibrium of the uranium radioactive series members. Indeed, the uranium enter into speleothem mainly as UO^{2+} soluble ions and is fixed into carbonate matrix where start to disintegrate so that older the speleothem age, higher the content of uranium daughter products. In this way, the radioactive equilibrium can be attained at a time higher than the speleothem age. This peculiarity should be taken into account when the speleothem age is calculated by considering the annual dose rate of uranium as the equilibrium one (Table 3.1).

In this regard, it is worth mentioning that not all the time the ESR lines in natural carbonates could be univocally attributed to some radiation free radicals, so that the most appropriate ESR lines are rather known by the numerical values of corresponding g-factors. Such is the case of resonance ESR lines having one of the following g-factor: 2.0058, 2.0032, 2.0020, 2.0014, 2.0012, 2.004 and 1.9976. For this reason, only by a detailed examination of the ESR spectrum of carbonate samples as well as the saturation behaviour of evidenced resonance lines it is possible to choose the most suitable ones.

At the same time, to increase the accuracy in evidencing and calculating the g-factors of different radiation radicals, beside using low-temperature ESR up to liquid nitrogen and helium, it is recommended to repeat the same measurements using higher frequencies such as K- and Q-bands, i.e. 18–27 and 35–50 GHz, respectively.

This approach, although not all the time available as the majority of actual spectrometers works in X-band, could significantly increase the measurement accuracy as it allows a better separation of the useful ESR lines (Fig. 3.5).

This observation suggests a certain degree of precaution when the amplitude of the maximum EPR signal is considered as a sole descriptor of radiation radicals' content. It is true that the amplitude of the EPR line is proportional to the content of the radiation free radicals, but in some cases, a single EPR line represents a superposition of few lines belonging to different radicals of which activation energy and consequently the accumulation rate could be different. In such case, the curve signal amplitude dose does not represent a superposition of exponential saturation curves which could lead to erroneous ages.

In this case and in similar ones, it is recommended either to repeat determinations by using higher frequencies (K, Q, W or ever higher) or to perform an isochronous annealing and to represent the ESR spectrum amplitudes by means of an Arrhenius plot. This procedure is very useful to determine the number of different species of radiation radicals of which ESR are partially or total superposed.

Regardless of these considerations, the ESR geochronology was successfully used to determine the absolute ages of corals, bivalve mollusc shells, planktonic foraminifera, speleothem, travertine and loess growing caliche. Corals, and at a lesser extent mollusc shells, represent, from this point of view, an exception as,

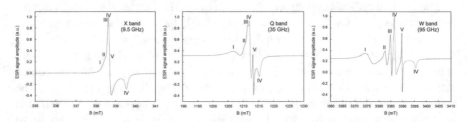

Fig. 3.5 The simulated ESR spectra of the complex line attributed to the orthorhombic CO_3 characterized by $g_\perp = 1.997$ and $g_\parallel = 2.0018$ (red colour spectrum). In fact, by passing to higher frequencies, this spectrum showed to be a superposition of at least five lines, four of them e.g. I, II, III, and V being symmetric and characterized by g-factors equal to 2.0020, 2.0014, 2.0012, and 2.004. Spectra simulated by EasySpin software

due to the fact that the calcium carbonate is produced and deposited at the interface between coral body and substrate, it represents a closed system less influenced by the ambiental variation of the uranium content of the sea water which chemical composition is almost stationary for long periods of time.

Phosphates Biogenic phosphates as CO_3-rich hydroxyapatite-dahllite with the empirical formula $Ca_5(PO_4)_{2.5}(CO_3)_{0.5}OH$ represents the main mineral components of the vertebrate bones as well as the mammals' tooth enamel. Tooth enamel consists of a multitude of hydroxyapatite micro-crystals linked together by a proteic 'glue' so it contains up to 96% hydroxyapatite making it an ideal material for ESR geochronology.

Given the relative uniformity of biogenic hydroxyapatite (dahllite), in almost all cases, the radiation radicals display a characteristic resonance line with an apparent powder pattern closed to a local orthorhombic radiation centre characterized by a $g_\perp = 1.997$ and $g_\parallel = 2.0018$. The resonance line at $g_\parallel = 2.0018$ showed to be the most reliable for age determination, although the X-band spectrum represents in fact a superposition of at least five different spectra as Q- and especially W-band measurements have suggested (Fig. 3.5).

This spectrum, previously identified in irradiated carbonates, is attributed to the carbonate anion CO_3 which substitutes a phosphate cation PO_4 in apatite to create the mineral dahllite, the main mineral component of the vertebrate bones.

In the case of tooth enamel ESR geochronology, as well as in the most general case of vertebrate bone fossils, a major source of uncertainties consists in uranium uptake. Uranium, due its solubility as a hexavalent uranyl ion, can be easily accumulated or washed out from tooth enamel. To overpass this situation, a more detailed investigation of uranium daughter products is necessary.

In this regard, it should take into account that thorium is completely insoluble so that the presence of ^{230}Th and ^{234}U could be used in inferring the time evolution of U uptake.

The main advantage of this approach consists in the possibility to check the results of the ESR dating with those of U/Th disequilibrium geochronology. Moreover, trying to adjust the ESR ages to the U/Th ones it makes possible to choose among different models of the uranium uptake the most appropriate one able to give similar ages.

Because the lifetime of the $g = 2.0018$ signal is estimated to be greater than 100 Ma, it results in that theoretically the maximum age which could be determined with a reasonable uncertainty is about 20–25 Ma which makes ESR geochronology compatible with K-Ar or Ar-Ar method.

As mentioned before, the ionizing radiation emitted by the natural elements existing in soil have different paths which means that in the case of thick fragments of fossils such as mammals' bones, of which dimensions are comparable with the main path of gamma rays, the attenuation effects should be taken into account, especially when the soil did not fill the bone cavities.

Quartz Quartz is the most abundant mineral on the Earth. Its presence is associated with volcanic rocks, as quartz is one of the main components of acid, granitic rocks, but can be found in significant proportions in metamorphic as well as sedimentary rocks.

Although, during crystallization of primary volcanic material, quartz tends to remove other elements which could contaminate its crystalline lattice, some elements such as aluminium, titan or germanium still are trapped, and, under the action of ionizing radiation generate radiation centres, some of them have a presumed half-life time in the order of 10^6 years.

The aluminium and titan centres display a complex ESR super hyperfine structure due to the interaction of the unpaired electron with the magnetic momenta of the neighbouring nuclei. Both spectra can be recorded at liquid nitrogen and lower temperatures due the relaxation effects. The other radiation centre such as germanium one, oxygen holes or E' centres can be observed at room temperature (Fig. 3.6).

Moreover, quartz is very resilient to weathering, so its presence in many sedimentary environments recommends quartz detritus as a potential medium for ESR geochronology.

A general property of all radiation centres consists in high-temperature insta-bility. When temperature increases, the amplitude of vibration increases which determines a recombination of radiation free radicals. Depending of the nature of radiation radicals, this process takes place at temperatures higher than 150–200 °C

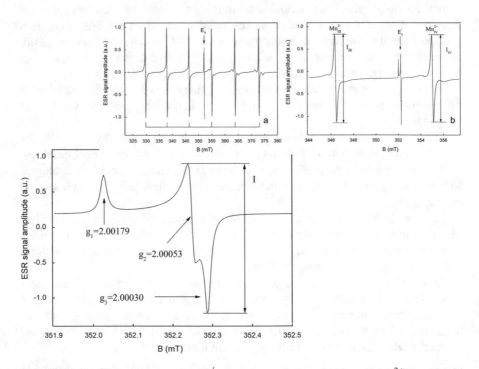

Fig. 3.6 The simulated ESR spectra of E'_1 centre in quartz superposed on the Mn^{2+} ions in CaO (insets a and b), the last ones used as reference for both g-factor and spectrum amplitude of the E'_1 centre

which resets the radiometric clock. This property was used to determine the age recent volcanic rocks such as pumice, tephra or rhyolite as well as of xenolithic quartz, as due to temperatures higher than 1000 °C, the radiation free radicals begin to accumulate when the quartz temperature is well below the resting one. As the volcanic material contains enough potassium to use the K-Ar and Ar-Ar radiometric age methods, it is possible to check the precision of ESR chronometry with the K-Ar and Ar-Ar, provided that the ages are of the order of thousands of kiloamperes.

The same peculiarity was used to determine the age of faults, as, due to the rock crushing during the fault time, the temperature increases sufficiently to anneal the radiation radicals accumulated in quartz. As in other cases, the ages such determined refer to the last fault event.

Case Study: ESR Dating of Quartz

In the next section, we apply the ESR dating methodology to identify the age of some probes of archaeological interest. For our study, we use the EasySpin simulation software package to generate the ESR spectra, while the paramagnetic parameters were selected from literature. One of the most commonly used systems in dating is the natural quartz which present the two possible paramagnetic centres generated by the ionizing radiation: i.e.(i) intrinsic defects, e.g. E_1' centre, peroxy centre, etc. and (ii) impurity-associated defects, e.g. Al centre, Ti centre, etc.

One of the great advantages of the ESR spectroscopy over the other dating techniques which use the effect of the ionizing radiation, e.g. thermoluminescence (TL) and optically stimulated luminescence (OSL), is that each type of defects has its own spectroscopic fingerprint given by the value of the paramagnetic parameters, i.e. the g-factor, the hyperfine constant or any other parameter used to describe the paramagnetic centre.

Even if the differences between those constants are small and the ESR signal can be overlapped, this can be changed by changing the working frequency *n.b.* today are available commercial ESR spectrometers with the working frequency between 1 and 263 GHz. So, by using appropriate experimental ESR condition, i.e. the working frequency or experimental temperature, one can select a particular paramagnetic centre on which the dating study can be conducted.

In the beginning, we consider in our analysis the intrinsic defects known in literature under the name of E_1' centre. In Fig. 3.6, one can see the simulation signal of E_1' centre in quartz using the literature value of the g-factor. All the values were obtained considering the working frequency of 9.5 GHz and the room temperature.

As a first step in the ESR dating, one can use an external standard sample. Because the g-factor value of the E_1' is close to the g-factor of the free electron, $g_e = 2.0023$, one of the best standards which can be used is the Mn^{2+} ion in a polycrystalline, cubic symmetry matrix, such as quicklime or CaO. With an electronic spin $S = 5/2$ and a nuclear spin $I = 5/2$, the Mn^{2+} ion in calcium oxide CaO shows a typical six hyperfine structure ESR lines (see Fig. 3.6).

In this case, the ESR intensity of the E_1' signal in quartz is normalized to the average of the third and fourth lines of the Mn^{2+} by using the following relation:

$$I(D_{abs}) = \frac{I_{E_1'}}{\left(\dfrac{I_{Mn_{III}^{2+}} + I_{Mn_{IV}^{2+}}}{2}\right)} \tag{3.4}$$

where we have used the intensities as reported in Fig. 3.6. Because, during the experiments, the standard sample is always in the same position inside the resonance cavity, the normalized signal of the ESR intensity improves the precision with which we determine the age.

For our study, we apply the *additive dose method* for two samples which show two typically different dependence on the additive dose due to the laboratory irradiation:

(i) one which shows a linear dependence of the ESR signal intensity with the additive dose. As one can see from the right site of Fig. 3.7, the unknown value of the paleodose, PD, can be obtained from the linear fit of the normalized ESR signal, $I_{ESR}(D_{abs})$, given by the following expression:

$$I(D_{abs}) = I_0\left(1 + \frac{D_{abs}}{PD}\right) \tag{3.5}$$

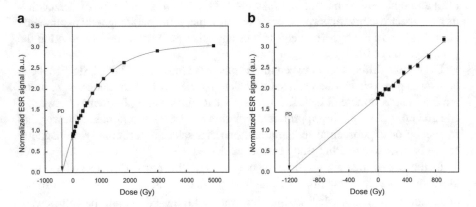

Fig. 3.7 Two examples of the use of the additive method in ESR geochronology. The growth curve of the ESR signal is simulated by an exponential saturation dependence (**a**) or by a linear dependency (**b**). In both cases, the PD is equal to the absolute value of the intercept. To illustrate the influence of experimental uncertainties, an error of 0.5% for the dose value and 2% for the normalized ESR signals were considered during fitting

where I_0 is the natural ESR intensity of the E_1' signal, i.e. the ESR signal due to the paleodose, before the laboratory irradiation and PD is the paleodose. From the best-fit parameters of the data from the right site of Fig. 3.7, one obtains the following values: $I_0 = 2.298 \pm 0.025$ and $PD = 1211.734 \pm 50.460\,Gy$

(ii) one which shows a saturation curve of the ESR signal intensity with the additive dose (see the left side of Fig. 3.7). The normalized ESR signal $I_{ESR}(D_{abs})$ is fitted with a saturation curve expression:

$$I(D_{abs}) = I_{sat}\left[1 - \exp\left(-\frac{D_{abs} + PD}{D_0}\right)\right] \tag{3.6}$$

where PD represents the paleodose, $I(D_{abs})$ represents the intensity of ESR signal for an absorbed dose D_{abs}, I_{sat} represents the ESR signal intensity for an infinite dose (saturation intensity) and D_0 represents the dose necessary for line intensity to be equal to $(1 - 1/e) \cdot I_{sat}$, $e = 2.718281\ldots$

From the best-fit parameters of the data from the right site of Fig. 3.7, one obtains the following values: $I_s = 3.053 \pm 0.016$, $PD = 384.036 \pm 8.321\,Gy$ and $D_{sat} = 1115.650 \pm 19.986\,Gy$

It is quite clear that the paleodose, PD, depends on the annual dose rate of natural radiation, D, and the time passed from the moment at which the paramagnetic centres began to accumulate in the quartz sample. The relation is straightforward:

$$T[y] = \frac{PD}{D}\,\frac{mGy}{mG/y} \tag{3.7}$$

From the analysis of the ESR signal as a function of the laboratory additive dose, the paleodose value is obtained in a direct way from the best-fitting parameters. However, the value of the annual dose rate is a subject to a much more imprecise. Often, the value of annual dose rate is considered the one obtained from Table 3.1, which is also the value used in our case in order to determine the ages of our samples from Eq. 3.7. However, in many real cases, this value could be quite different, and this will result in a different age than the real one: i. - if annual dose rate is overestimated, the age value will be smaller than the real one, ii. - if the annual dose rate is sub-estimated, the sample will appear older than the real one.

In this regard, it is worth remarking that the values of the annual dose debt are always calculated considering the present composition of neighbouring medium. If during the sample conservation the distribution of the main radioactive elements fluctuated by different causes, it is obvious that the precision of age determination will be affected. This fact could be viewed as one of the explanations of the reduced precision of the ESR geochronology, but regardless of this remark, ESR dating remains one of the main absolute geochronological methods, especially in the case of Upper and Middle Pleistocene ages (11.7 to 774 ka).

For our case study, we have considered the value of the annual dose rate the value obtained from Table 3.1, respectively, $D = 1.0473 \times 10^{-3} + 0.8131 \times 10^{-3} +$

$3.0523 \times 10^{-3} = 4.9127 \times 10^{-3}$ Gy/y . The last step in our case is to determine the age by applying Eq. 3.7 for the two samples using the paleodose value obtained from the best-fit parameters. For the sample which shows the linear dependence, one obtains:

$$T[y] = \frac{1211.730}{4.9130 \times 10^{-3}} \frac{Gy}{Gy/y} \Rightarrow T = (246650 \pm 10270) \, y \qquad (3.8)$$

and for the sample with the saturation curve dependence, one obtains:

$$T[y] = \frac{384.040}{4.9130 \times 10^{-3}} \frac{Gy}{Gy/y} \Rightarrow T = (78170 \pm 1670) \, y \qquad (3.9)$$

New Developments in Quartz Dating

In the last decade, the ESR dating shows new opportunities, specially related to the application of the regenerative dose method, applied to the second type of irradiation effects, i.e. impurity-associated defects: Al and Ti centres (see Fig. 3.8). In the literature, it has been reported different steps in the SAR protocol for ESR

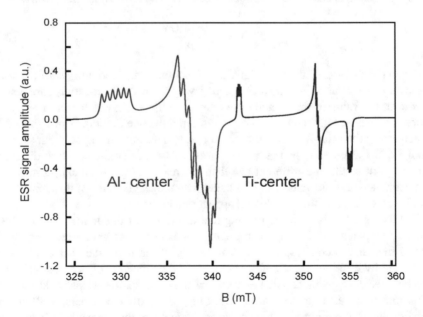

Fig. 3.8 The simulated ESR spectra of Al and Ti ions in quartz. For the simulation, we have considered only the Al centre, $[AlO_4]^0$, in interaction with nucleus ^{27}Al and Ti centre, $[TiO_4/Li^+]^0$, in interaction with nucleus ^7Li. The paramagnetic parameters were used from literature and the temperature was $T = 100$ K

dating; however, this is still a subject of debate and the protocols are the subject of future research.

However, there are some important aspects which appear in all these new developments. First of all, in the ESR dating, the impurity-associated defects and not intrinsic defects are used. Due to the relaxation effect which brought very much the ESR lines, the ESR dating experiment on Al or Ti should be carried out at low temperature, usually around 100 K ($-173°C$). It is worth mentioning here that the new results demonstrate that using Al or Ti centres one obtains the same value of paleodose by using appropriate steps in the protocols, e.g. preheating treatment to the irradiated aliquots. Therefore, preheating treatment is a common procedure of any ESR dating protocol. To understand the preheating treatment, we need to look in a deeper way to the generation of the ESR centres used for dating.

The only one limitation in the upper dating limit is due to the thermal lifetimes of the Al and Ti centres which for quartz samples obtained the value of ≈ 1.7 Ma for Ti and ≈ 1.5 Ma for Al at 10°C. To reach these limits, one needs to include in the protocol a preheating treatment. To understand the need of this treatment, one must understand the physical process for the impurity-associated defects' production. As was mentioned in section "ESR Age Determination", under the influence of ionizing radiation, the electrons or the holes are trapped by impurities given by ESR signal associated with Al^{3+} or Ti^{4+} ions in quartz. Therefore, the ESR signal is given by the number of ions with trapped electrons/holes, and this number is proportional to the natural dose rate. Even if the traps associated with one ion have the same ESR signal, their depth is not the same for all ions. In order to record the ESR signal given only by those electrons or holes found in the deepest traps characterized by the thermal lifetimes of the Al or Ti ions, one must ensure that all the other paramagnetic centres are cancelled, and this is possible by performing different preheating treatments.

However, the new development in quartz ESR dating is a field of continued challenges for established accepted protocols for all scientific community.

Concluding Remarks

The ESR geochronology belongs to the absolute dating techniques based on the accumulation of radiation damages produced by natural radioactive sources, similar to thermoluminescence (TL) and optically stimulating luminescence (OSL). Different from these, the ESR has a supplementary capacity to increase its sensitivity and resolution by using higher frequency than the usual X-band or by performing measurements at liquid helium temperature. As some radiation centres have a lifetime of up to few milliamperes, the ESR geochronology covers an age interval which corresponds to Holocene and Upper Pleistocene, partially covered by TL, OSL and uranium disequilibrium (UD) methods.

With respect to TL, OSL and UD, ESR remarks itself by a more simplified measured protocol, without tedious analytical procedure such as in the case of UD or the necessity to prepare samples in almost absolute darkness as in the case of OSL.

On the other hand, the ESR precision is strongly influenced by the long-term fluctuation of the environmental radioactivity which could induce systematic errors up to 15–20%. For this reason, it is recommendable to check anytime possible the ESR ages with the ages furnished by alternative geochronological methods, if such methods are available.

But regardless of these inconveniences, the ESR geochronology remains one of the most appropriate dating techniques for a great variety of objects of which ages vary between few kiloamperes and more than one milliampere.

Quiz

1. What is an absolute geochronology method?
2. What is the principal difference between ^{14}C and ESR absolute geochronology?
3. Which is the role of natural radioactive elements for ESR geochronology?
4. What are the differences between the ESR reconstructive and regenerative dose methods?
5. Enumerate some advantages of ESR dating vs. TL/OSL dating.
6. Using data from Fig. 3.4, calculate the age of the sample.

Acknowledgments This work was supported by a grant of the Romanian Ministry of Research and Innovation, CCCDI UEFISCDI, project number PN-III-P1-1.2-PCCDI-2017-0686/52PCCDI/2018, within PNCDI III, by the Romanian Ministry of Research and Innovation, project number 15PFE/2018, and partially realized with the Cooperation Protocol No. 4322-4-2020/2022 between Joint Institute for Nuclear Research, Dubna, Russian Federation and University of Bucharest.

References

W.E.H. Blum, P. Schad, S. Nortcliff (2018) Essential of Soil Science, Borntraeger Science Publischers, Suttgart, ISBN 978-3-443-01090-4

P-C. Zhang, P.V. Brady (Eds.) (2002) Geochemistry of Soil, Radionuclides, Soil Science Society of America Special Publication Number 59, Soil Science Society of America, Madison, Wisconsin

G.B. Dalrymple (2001) The age of the Earth in the twentieth century: a problem (mostly) solved, Geological Society, London, Special Publications, volume 190, https://doi.org/10.1144/GSL.SP.2001.190.01.14

G. Gurin, N. Mercier, G. Adamiec (2011) Dose-rate conversion factors: update, Ancient TL, 29, 5–7

M. Ikeya (1993) New Applications of Electron Spin Resonance, World Scientific Publishing, Singapore, ISBN 981-02-1199-6

W.J. Rink (1997) Electron Spin Resonance (ESR) Dating and ESR Applications in Quaternary Science and Archaeometry, Radiation Measurements 27, 975–1025

S. Tsukamoto, S. Toyoda, A. Tani, F. Oppermann (2015) Single aliquot regenerative dose method for ESR dating using X-ray irradiation and preheat, Radiation Measurements, 81, 9–15.

S. Stoll, A. Schweiger (2006) EasySpin, a comprehensive software package for spectral simulation and analysis in EPR, J. Magn. Reson. 178, 42–55.

M. Richter, S. Tsukamoto, H. Long (2020) ESR dating of Chinese loess using the quartz Ti centre: A comparison with independent age control, Quaternary International, in press.

S. Tsukamoto, N.Porat, C. Ankjrgaard (2017) Dose recovery and residual dose of quartz ESR signals using modern sediments: Implications for single aliquot ESR dating, Radiation Measurements, 106, 472–476.

M. Asagoe, S. Toyoda, P. Voinchet, C. Falgures, H. Tissoux, T. Suzuki, D. Banerjee (2011) ESR dating of tephra with dose recovery test for impurity centers in quartz, Quaternary International, 246, 118–123.

W.J. Rink, J. Bartoll, H.P. Schwarcz, P. Shane, O. Bar-Yosef (2007) Testing the reliability of ESR dating of optically exposed buried quartz sediments. Radiation Measurements, 42, 1618–1626.

A.S. Murray, A.G. Wintle (2000) Luminescence dating of quartz using an improved single-aliquot regenerative-dose protocol. Radiation Measurements, 32, 57–73.

A.S. Murray, J.M. Olley (2002) Precision and accuracy in the optically stimulated luminescence dating of sedimentary quartz: a status review. Geochronometria, 21, 1–16.

S. Tsukamoto, H. Long, M. Richter, Y. Li, G.E. King, Z. He, L. Yang, J. Zhang, R. Lambert (2018) Quartz natural and laboratory ESR dose response curves: A first attempt from Chinese loess, Radiation Measurements. 120, 137–1142

Chapter 4
ESR Applications in Food Science

Free Radicals in Food Science

Food science can be considered a boundary discipline between chemistry, physics, agricultural science, nutrition and alimentary hygiene. For this reason, the food science is at the same time fundamental and applied. It deals not only with the nature of main components which enter in the composition of our daily food but also with the technologies and processes to produce and conserve it as well as with a thoroughly quality assessment. In the food composition enter a great diversity of vegetal, animal as well as mineral components, including more organic synthetic compounds utilized to increase the shelf life together with colour, aroma or taste.

At present, the global production of food overpasses 10 billion tons, which poses serious problems concerning proper storage, seldom difficult to be achieved. This generates a complex problem concerning long-time conservation and quality control.

By the end of the ninth century, Louis Pasteur discovered and promoted the food conservation procedure, known today as pasteurization, based on the inactivation of spoiling microorganisms by exposing canned food to high temperature. Pasteurization replaced the older processing by salting or fumigation which significantly altered the organoleptic properties of conserved food. With the advent of nuclear physics, the bactericidal effects of high-energy ionizing radiation such as X – or gamma – ray were evidenced and systematically investigated. Beginning with the second half of the twentieth century, the food decontamination by exposing to high-energy ^{60}Co gamma-rays became a routine technique. In view of the high energy transported by nuclear radiation, one of the first results of this procedure is the generation of free radicals (FR) whose content monotonously increased with the amount of energy absorbed from the gamma-ray radiation field. As any FR contains at least one unpaired electron, the electron spin resonance (ESR) spectroscopy represents the method of choice to evidence their presence and concentration. For

© Springer Nature Switzerland AG 2021
A. K. Shukla, *ESR Spectroscopy for Life Science Applications: An Introduction*,
Techniques in Life Science and Biomedicine for the Non-Expert,
https://doi.org/10.1007/978-3-030-64198-6_4

this reason, the FR ESR signal consists of one or few resonance lines whose g-factor is close to the free electron gyromagnetic factor, i.e. 2.0021. This peculiarity represents an advantage as the existence of radiation FR can be easy identified by their ESR spectrum. On the other hand, FR spectra consist of a reduced number of lines whose g-factor is close to the free electron one. For this reason, their true nature cannot be unequivocally ascertained.

It should be pointed out that in the case of ESR, g-factor has a rather phenomeno-logical character. It is equal to the ratio between the energy of the microwave field and the corresponding resonance magnetic field of a given ESR line:

$$g = \frac{h \cdot \nu}{\beta \cdot B} \tag{4.1}$$

where ν is the frequency of electromagnetic field, h is Planck's constant, β is Bohr's magneton and B is the resonance magnetic field.

The g-factor should not be confounded with the gyromagnetic factor γ which relates magnetic momentum μ and angular momentum J of an electron or a nucleus:

$$\mu = -\gamma \cdot J \tag{4.2}$$

Accordingly, the g-factor is strongly influenced by the strengths and the symme-try of local electric field as well as by the hyperfine interaction with the magnetic momentum of neighbouring nuclei.

It is worth mentioning that the FR are not the only results of food irradiation. Besides FR, high-energy ionizing radiation can modify the fat structure; induce thermoluminescence (TL) centres in the mineral impurities, accidentally entered into food; determine DNA alteration; or do not sterilize completely food so a small fraction of microorganism survived radiation decontamination.

Regarding TL, it is worth mentioning that this type of luminescence is specific of a large category of crystalline materials including minerals, which, after absorbing energy from ionizing radiation, re-emit it as light when heated. Similar to ESR, the amplitude of TL signal, i.e. the number of emitted photons, is positively correlated with the amount of absorbed energy.

To standardize the assessment of food irradiation, the European Union has adopted a set of procedure to be used for the screening of irradiated food, three of them devoted to use the ESR to evidence the irradiation FR (Table 4.1).

In foods, as well as in a multitude of organic compounds, FR are usually generated under the action of external factors, ionizing radiation or the ambient oxygen being some of them. FR, in view of their high reactivity, are unstable entities whose life-time varies between femto- and milliseconds. This makes their investigation by continuous-wave (CW) ESR almost impossible in liquid or viscous media. On the contrary, the FR generated under the action of ionizing radiation in solid or dehydrated substances have a life-time up to few years, so their presence can be evidenced by ESR without major problems.

Table 4.1 The European Union Committee for Standardization procedures to detect the irradiated foods. Bold fonts evidence the ESR-based protocols

Code	Title
EN1784:2003	Detection of irradiated food containing fat. GC analysis of hydrocarbons
EN1785:2003	Detection of irradiated food containing fat. GC/MS analysis of 2-ACBs
EN1786:1997	**Detection of irradiated food containing bone. Method by ESR spectroscopy**
EN1787:2000	**Detection of irradiated food containing cellulose by ESR spectroscopy**
EN1788:2001	TL detection of irradiated food from which silicate minerals can be isolated
EN13708:2002	**Detection of irradiated food containing crystalline sugar by ESR spectroscopy**
EN 13751:2009	Detection of irradiated food using photostimulated luminescence
EN13783:2002	Detection of irradiated food using direct epifluorescent filter technique/aerobic plate count
EN13784:2002	DNA comet assay for the detection of irradiated foodstuffs. Screening method
EN14569:2004	Microbiological screening for irradiated food using LAL/GNB procedures

In checking the food quality, currently the ESR is used in two main directions: -i. to evidence any sterilization by gamma-ray or in few instances by high-energy electrons and -ii. to assess the oxidative processes. Although the same spectroscopic method is used in both situations, there are some essential differences between them.

In view of their behaviour, FR can be either indirectly evidenced by means of spin trapping (ST) or spin labelling (SL) as in the case of food component rich in water or lipids or directly, as in the case of solid component.

Spin Trapping

Spin trapping is an experimental technique utilized to detect FR when their half-life is too short to be evidenced directly by CW ESR. In this case, to detect and quantify their presence, there are specific reactive, the spin traps (ST), which by interacting covalent, in aqueous or lipid media, with FR generate new stable FR called adducts whose presence can be evidenced and investigated by ESR. To be evidenced by ESR, ST concentration should be around 10^{-4} to 10^{-6} Mole.

ST are diamagnetic compounds which belong to the nitrone as well as nitroso class of organic chemistry. Nitrone ST consist of an N-oxide and an imine with the general structure $R_1R_2C=N^+R_3O^-$ characterized by a nitrogen-carbon double bond. Nitroso class has a NO group attached to an organic moiety with the general formula $O=N-CR_1R_2R_3$. Nitrone ST can have a linear or a cyclic structure (Fig. 4.1). On the contrary, nitroso ST, due to their configuration, can have only linear structure (Fig. 4.2). The difference between nitroso and nitrone ST is reflected

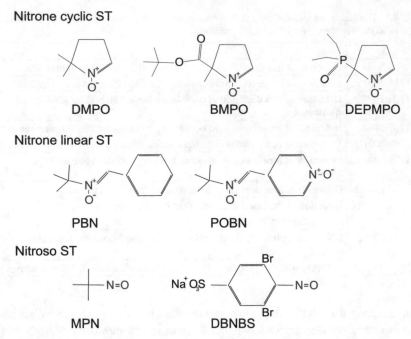

Fig. 4.1 Structural formula of some of the most utilized ST. DMPO, 5,5-dimethyl-1-pyrroline-N-oxide; BMPO, 5-tert-butoxycarbonyl-5-methyl-1-pyrroline-N-oxide; PB, N-tert-buthyl–phenylnitrone; DEPMPO 2-diethylphosphono-2-methyl-3,4-dihydro-2H-pyrrole 1-oxide; PBN, N-tert-butyl-α-phenylnitrone; POBN, α-(4-pyridyl 1-oxide)-N-*tert*-butylnitrone; MPN, 2-methyl-2-nitrosopropane; DBNBS, 3,5-dibromo-4-nitrosobenzenesulfonic acid sodium salt

Fig. 4.2 Reaction of a nitrone ST (**a**) and of a nitroso ST (**b**) with a FR to generate nitroxide radical adducts

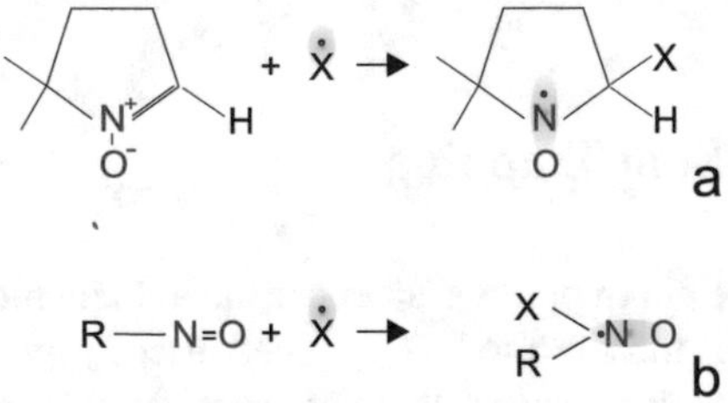

by the adduct structure (Fig. 4.2) and, consequently, by the presence of hyperfine and superhyperfine structure in their ESR spectrum.

In the process of adduction, the FR unpaired electron is transferred to nitrogen atom of the nitrone or nitroso ST making the spin adduct, the new FR, significantly more stable than the captured one. Therefore, the ESR spectrum of adduct is strongly influenced by the nature of captured radical. Figure 4.3 illustrates the different superhyperfine structures thus generated for the α-nitrogen hyperfine triplet (Fig. 4.3).

In the case of nitrone ST, the external FR are trapped by a β-carbon with respect to aminoxyl nitrogen so that the unpaired electron of the trapped radical is localized

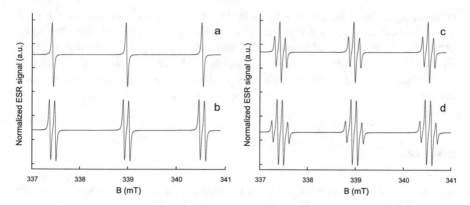

Fig. 4.3 X-band typical ESR spectra of nitroso ST adducts consisting one unpaired electron (**a**), a - CH-trapped moiety (**b**), a -CH_2CH_3 moiety (**c**) and a -CH_3 moiety (3d). (A_N = 1.55 mT, A_H=0.122 mT)

around α-nitrogen (Fig. 4.2a). Thus, the final ESR spectrum consists of a nitrogen triplet, each line being split into two superhyperfine lines by the β-hydrogen located on the captured radical (Fig. 4.2a).

Contrarily, the nitroso FR adducts are formed by bonding the environmental FR to nitrogen atom which enable direct interaction of the added radical with the nitrogen nucleus. This gives rise to a complex ESR spectrum which better reflects the structure of captured radicals (Fig. 4.2b). This property is illustrated by the ESR spectra of nitroso adducts formed when an unpaired electron of nitroxide radical interacts with the nuclear spin of the adjacent imminoxyl nitrogen (Fig. 4.3a), with the β-hydrogen of a -CH-trapped moiety (Fig. 4.3b), with the β-hydrogen of the -CH_2CH_3 moiety (Fig. 4.3c) as well as with the β-hydrogen nuclei of a -CH_3 moiety (Fig. 4.3d).

An ideal ST should generate adducts with a great diversity of FR concerning both chemical nature and time life. Accordingly, it is expected that the adduct life time should be significantly greater than those of captured FR, while the adduct ESR spectrum should permit a rapid identification of captured FR.

Oxidative processes in foods investigated by ESR and ST Long-term stability of foods is all the time threatened by the action of environmental oxygen which makes them improper for consumption. This process can be slowed down by natural antioxidants existing in food whose action can be evidenced by the ESR spectra of trapped FR. The most common types of oxidation FR are neutral OH and hydroxyl -OH. Depending of the type of food, they can interact with ethanol, in the case of beverages or with fatty acids in the animal or vegetal fats. For this reason, ST are used to assess the antioxidant potential of different compounds such as thiols, ascorbic acid, glutathionic acid, etc.

Oxidation processes can be monitored by following the time evolution of ESR signal amplitude of the adduct under different storage or thermal treatment conditions, in the presence as well as in the absence of antioxidants. In the most cases, the oxidative stress was induced by increasing the storage temperature or by prolonged exposure, up to few months, to atmospheric oxygen. In all cases, the ESR signal adduct amplitude shows a typical time dependence which tends to a saturation exponential function as the available ST molecules are gradually converted into adducts.

Spin Labelling

Opposite to ST, spin labelling (SL) is another experimental technique, currently used to investigate the structure and processes at molecular scale, such as cell membranes, water molecules' interaction with proteins' side chains, oxidative stress, antioxidants' kinetics, etc.

SL are stable paramagnetic organic compounds with a nitroxide radical whose unpaired electron is delocalized on the π orbitals of the N-O bond and steric stabilized by neighbouring carbon or methyl group (Fig. 4.4) whose ESR spectrum consists of a hyperfine structure singlet (Fig. 4.5a). Only in the case of 2,2-diphenyl-1-picrylhydrazyl (DPPH), the N-O bond is replaced by a N-N one (Fig. 4.4c) resulting in a more complex ESR spectrum. This one consists of five lines due to free electron delocalization with equal probability on two nitrogen atoms (Figs. 4.4c and 4.5b).

SL application in food science The main application of SL in food science, besides fundamental issues concerning the molecular structure of aliment different constituents, is related to the investigation of antioxidants' capacity of different

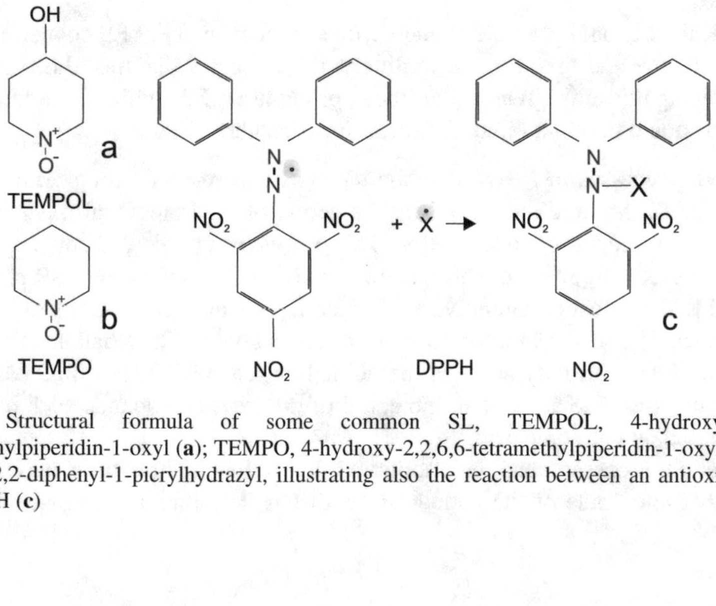

Fig. 4.4 Structural formula of some common SL, TEMPOL, 4-hydroxy-2,2,6,6-tetramethylpiperidin-1-oxyl (**a**); TEMPO, 4-hydroxy-2,2,6,6-tetramethylpiperidin-1-oxyl (**b**); and DPPH, 2,2-diphenyl-1-picrylhydrazyl, illustrating also the reaction between an antioxidant and the DPPH (**c**)

natural organic compounds, i.e. natural antioxidants. Accordingly, their role consists of to reduce metabolic FR.

In food science, antioxidants are an important class of additives that react with the oxygen even in frozen or refrigerated food by reducing the spoilage caused by oxidants such as colour modification or fat rancidity. The NA contains especially polyphenols and carotenoids exhibiting a wide range of beneficial biological effects. At present, there are in use more natural and synthetic antioxidants. Ascorbic acid (E300) and tocopherols (E306-E309) belong to the first category, while propyl gallate (E310), butylated hydroxyanisole (E320) and butylated hydroxytoluene (E321) enter in the second category of food antioxidants, although the total number of antioxidants overpasses 250.

It should be remarked that NA interact directly with FR generated during natural metabolic processes or by the action of external factors such as ionizing radiation (ultraviolet, X and gamma rays) reducing them in a reaction one-to-one, i.e. one molecule of antioxidant can reduce only one FR. As SL are in fact FR, they are neutralized by antioxidant molecules, so that their concentration decreases over time. This fact gives a quantitative evaluation of the capacity of antioxidants to act as FR scavengers.

Consequently, the experimental procedure consists of tracking the gradual reduction of SL ESR signal which gives information on the reaction order and the ability of antioxidants to neutralize FR (Fig. 4.5). For this reason, spin labelling represents a remarkable experimental technique to investigate the dynamic properties of a large scale of complex organic compounds, but, in the case of FR, the information furnished by this technique is restrained only to evidence the presence as well as the time behaviour of FR, but not on the their structure, as in the case of nitroso ST.

The most utilized SL to evidence the antioxidant capacity of different agents are 2,2,6,6-tetramethyl-1-piperidinyloxy (TEMPO) and DPPH. Both SL are soluble in polar (TEMPO, TEMPOL) as well as nonpolar (TEMPO, TEMPOL, DPPH) solvents which facilitate the study of the antioxidant action in liquid phase. As mentioned before, the antioxidant activity can be estimated by measuring the rapidity with which the SL ESR signal diminishes until its complete extinction.

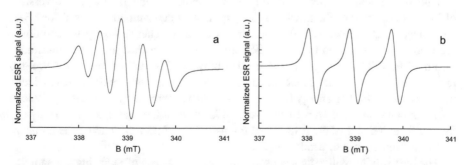

Fig. 4.5 X-band simulated ESR spectra of 2,2,6,6-tetramethyl-1-piperidinyloxy (TEMPO) (**a**) and SL 2,2-diphenyl-1-picrylhydrazyl (DPPH) (**b**)

It is worth mentioning that in some situation, the generation of FR in the case of ST or the activation of molecular centres capable of oxidizing SL can be triggered by external factors such UV or X-ray irradiation or as a result of some in situ chemical reaction. The actual ESR spectrometers permit all irradiations to be performed in the ESR spectrometer resonance cavity. This direct procedure is the most recommended for their accuracy and sensitivity.

ESR and Irradiated Food

The assessment of irradiated food can be considered one of the most common applications of the ESR spectroscopy in food science. As mentioned before, the decontamination of food by exposing to X-rays was among the first application of emerging nuclear science. Now, the high-energy food irradiation for conservation is a routine procedure. Depending on the specific task, irradiation can be done at doses between 1 kGy to kill insects in fruits, vegetables, grain or dried meat and 10 kGy and even above this threshold for a complete food sterilization. It is obvious that at these doses which are few orders of magnitude higher than the natural background, irradiation generates a significant amount of radiation FR in all parts of food.

At the same time, FR, regardless of their nature and how they are generated, are very reactive entities having the tendency to react with neighbouring molecules in ps and even in much shorter time. Exception from this rule are the hard part of food such as bones, shells, exocarp of seeds or any other hard part able to stabilize FR and prolong their life span up to month and even years.

Although the quantity of food which is decontaminated and conserved by irradiation represents only a small fraction of the total food, a correct customer information concerning any radiation treatment is compulsory. Given the great diversity of food products, beside ESR, there were elaborated more other methods to evidence any high-energy radiation treatment applied to increase the shelf life, to prolong the conservation period or even to sterilize completely a certain category of food.

Due to chemical complexity, the high-energy radiation generates a great diversity of FR, but only few of them, whose structure is well determined, are used for ESR identification. Accordingly, FR generated in bones or shells, sugar and cellulose are well known and used in routine ESR monitoring. Besides, there is a multitude of other FR which can also be used for this purpose. Sometimes, by irradiation, in the same sample, there are generated more FR whose g-factors are quite close which needs the use of higher-energy ESR (K-, Q- or W-band) to discriminate them. When only one band ESR spectrometer is used, an isochronal thermal treatment followed by an Arrhenius plot of the ESR signal intensity is useful in evidencing the number and activation energy of irradiation FR.

The best way to realize an isochronal annealing consists of exposing the sample for the same interval of time to gradually increasing temperature followed, for each temperature, by a rapid quenching in a mixture of water and ice and recording the existing ESR spectrum. By representing the natural logarithm of ESR spectrum amplitudes versus inverse of absolute temperature, the resulting curve permits to determine the number of FR species as well as their activation energy.

The recombination process of FR can be described by a first-order reaction which leads to an exponential decay function of the time dependency of the initial number of FR. This particularity can be experimentally evidenced by accelerated ageing of irradiated sample by isothermal annealing. Depending on how many different types of FR of which ESR lines are superposed are generated by irradiation, the decay curve of the FR ESR signal will correspond to one or more exponential decay curves which after deconvolution makes possible to establish the real number of different species of FR. In this regard, spectrum simulation represents one of the most straightforward procedures to check the accuracy of any model, given the multitude of specialized software, some of them being open-source and very friendly to use.

According to the European Committee for Standardization, there are three categories of food components which can be used to assess any previous food treatment with ionizing radiation: bones (fish, chicken, beef, pork, lamb), crystalline sugar and cellulose. In all cases, only the hard components can be used due to their capacity to retain FR for prolonged time.

In view of the reduced life-time of FR, ESR detection of any irradiation treatment applied to food, regardless of the type or FR structure, should be considered as a screening procedure in the sense that the presence of FR could indicate a previous irradiation, while the lack of any signal does not prove the irradiation was not done.

Bones and shells Bones, the main constituent of the hard inner skeleton of majority of vertebrates (lampreys, sharks and partially sturgeons lack bony skeleton), are composed of hydroxyapatite $Ca_{10}(PO_4)_6(OH)_2$ micro-crystals bound together by collagen and other proteins. The irradiated samples show different ESR signals generated by molecular ions from which the most studied is the CO_2^- radical. On the contrary, the egg shell as well as the exoskeleton of edible molluscs, such as oysters, scallops, mussels, land and marine snails and cuttlefish, contains calcium carbonate mainly as calcite $CaCO_3$, but which after irradiation shows an ESR signal very close to the CO_2^- radical characteristic for the irradiated hydroxyapatite.

The characteristic signal of the CO_2^- radical (Fig. 4.6a) consists of an asymmetric line suggesting an axial local symmetry. This line is characterized by two effective g-factors, 1.988 and 2.002, the first one corresponding to a perpendicular orientation of the magnetic axis of the $CaCO_2$ molecule radical with respect to external magnetic field and the second one perpendicular to the magnetic field orientation. However, a closer inspection shows that the ESR line of such natural sample presents a much more complex spectrum. In fact, this line is due to a superposition five lines, the axial one previously mentioned and another four almost isotropic having g-factors between 2.0014 and 2.0058. This peculiarity appears as a consequence of the complex crystalline structure of the hydroxyapatite which could present different environment symmetries, even for the same molecular radicals.

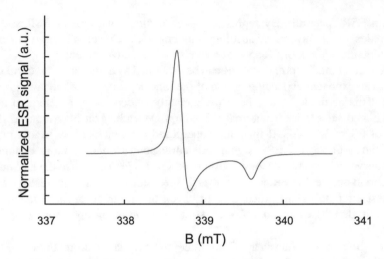

Fig. 4.6 X-band complex ESR spectrum of CO_2^- radical found in almost all hydroxyapatite bones and calcite invertebrate exoskeleton. Despite its apparent simplicity, this ESR signal represents in fact a superposition of ESR spectra of at least three different FR, whose g-factors are too close to be evidenced by X-band ESR

A general characteristic of the majority of radiation FR consists of the fact that they have g-factors whose values cover a very narrow interval between 1.988 and 2.003. This peculiarity makes their identification in X-band very difficult. To overcome this inconvenience, there are different approaches for an accurate analysis of the irradiated bones and shells. One of them is to analysis the ESR signal as a function of microwave power. This type of analysis is based on different relaxation mechanisms which characterized the coupling between the paramagnetic centres at the sounding environment, i.e. the time necessary to lose the absorbed energy. Different power saturation can be a sign of distinct paramagnetic centres in various environment symmetries. Another approach is to investigate the temperature stability of the free radical induced by irradiation. Moreover, if thermal stability is investigated by an isothermal annealing, one can obtain from the Arrhenius plot the activation energy for each ESR signal. Different activation energies can be due to distinct paramagnetic centres.

In the last decades, due to the high frequency-high field development of ESR spectrometer, one of the most appropriate approaches is the multi-frequency ESR experiments. They can cover the frequencies' range from 1 to 263 GHz which significantly increases the spectral resolution.

The sensitivity of actual ESR spectrometer allows, if the measurements are performed at room temperature, to evidence the presence of $CaCO_2$ at absorbed doses greater than 0.5 kGy in the case of irradiated bones or shells, i.e. the minimum dose necessary to prolong the shelf life of fresh products.

As the life span of the $CaCO_2$ FR can overpass 1 year in frozen bones and shells, it is possible, with a certain degree of approximation, to estimate the initial dose of the decontamination treatment. This can be done by the additive dose technique, i.e. the same fragment of hard tissue is irradiated with progressive doses, after each irradiation its ESR spectrum being recorded in exactly the same experimental conditions to minimize the accidental errors due to sample manipulation. Then, the signal intensities and corresponding added dose can, by using a least-square method, reconstruct the amplitude vs. dose characteristic linear dependency whose intercept represents the absorbed during irradiation. This can be used provided that the FR content remained unchanged. What is most important, its results should be considered rather informative in the absence of confident data concerning the long-term stability of irradiation FR.

This method is currently used to evidence any irradiation treatment applied to animal meat containing bones or fragment of bones as the case of mechanically deboned meat, edible shell, egg shell, etc. Given the high stability of $CaCO_2$ FR in bones, this procedure can even be applied to the culinary preparations.

Crustaceans such as lobsters, shrimps, crabs and freshwater crayfishes represent a well-appreciated source of proteins but, at the same time, due to the aquatic media where they live, could be contaminated with pathogenic germs belonging to *Salmonella, Shigella* or *Vibrio* genera. All crustaceans possess a hard exoskeleton, the carapace, which protect the animal soft body. The crustacean carapace consists of chitin, the N-acetylglucosamine polysaccharide and a significant amount of calcium carbonate. In majority of cases, the ESR spectrum of non-irradiated crustacean cuticle or carapace presents a weak sextet attributed to Mn^{2+} ions. When irradiated at doses up to $10\,kGy$, the cuticle displays a complex ESR spectrum located between the third and the fourth Mn^{2+} hyperfine sextets, whose presence could be observed after more than 1 year, assessing ESR suitability for irradiated edible crustaceans.

Fruits and vegetables The disaccharide sugar (fructose + glucose joined by a glucose bond) and the polysaccharide cellulose are the main compounds whose presence in irradiated products of plant origin can be identified by ESR spectroscopy. Polycrystalline sugar is present in a majority of dried fruits such as raisin, mangoes, paprika and so on. Sugar or sucrose forms monoclinic crystal which accumulates in the fleshy part of the fruit, in stem or in the root. It is worth remarking that only polycrystalline sugar present in any part of fruits or vegetables generates, after irradiation, a detectable specific ESR signal. Unirradiated sugar has no ESR signal, but after irradiation, it shows a relatively complex spectrum centred around a g-factor of 2.005 with a not well-resolved hyperfine structure consisting of nine lines, most probably due to the delocalization of an unpaired electron over eight hydrogen ions (Fig. 4.7).

Sometimes, even in the absence of irradiation, an ESR signal consists of a single line centred around a g-factor equal to 2.007 and depending on the plant nature is present. Most probably, this signal is due to semiquinone radicals present in majority of plant hard tissues.

Cellulose is the second major component which can be found in almost all plant tissue. When irradiated, the cellulose exhibits a specific ESR consisting of

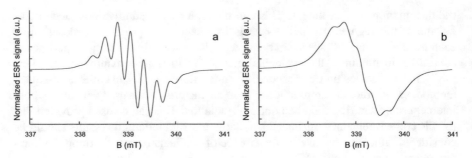

Fig. 4.7 X-band simulated ESR spectra of monocrystalline (**a**) and polycrystalline (**b**) sucrose

a hyperfine triplet centred around a g-factor equal to 2.004. As the amplitude of the central line is more than two times greater than the amplitude of satellite ones, it is possible that the irradiated cellulose central line is superposed onto the irradiated lignin line. The cellulose radiation FR showed to be relatively unstable, so the absence of its ESR signal does not means the food was not irradiated.

A special attention should be paid to the moisture content of sample as the presence of water contributes to the FR recombination. This precaution should be taken anytime when irradiated food is analysed for any radiation treatment.

A special attention should be paid to the moisture content of sample as the presence of water contributes to the FR recombination. This precaution should be taken anytime when irradiated food is analysed for any radiation treatment.

Mushrooms Mushrooms are living organisms which belong to kingdom Fungi, quit different from kingdoms Animalia and Plantae. Their cells walls have chitin, one of the main characteristics which differentiate mushrooms from the other living organisms. Edible mushrooms represent a well-appreciated aliment, consumed almost all over the world. Compared to meat, mushrooms have a lower nutritional value, mainly due to a reduced content of fat (~0.3%) and protein (~3%). At the same time, mushrooms are richer in riboflavin, niacin and pantothenic acid. Wild mushrooms, which are collected mainly in summer and autumn, are traditionally conserved by air-drying which exposes them to different spoiling microorganisms. Hence, to prolong their shelf life, irradiation with ionizing radiation appears as a good alternative.

When irradiated with gamma-ray, dried mushrooms show, depending on genera and species, different ESR spectra, mainly consisting of single line. In some cases, this line displays an unresolved hyperfine spectrum whose amplitude increased with the absorbed dose by following a saturation exponential dependence. Remarkable was the stability of radiation FR which in some cases persisted for more than 18 months.

Given the widespread consumption of mushrooms, a systematic future investigation of irradiation FR in mushrooms could be very promising.

Concluding Remarks

The food decontamination by exposing to high-energy ionizing radiation showed to be an effective method to prolong the shelf life of a diverse category of foods, including fruits, vegetables, seafood, meat of different origin, etc. Besides decontamination, radiation treatment induces in food a multitude of free radicals whose life-time strongly depends on the water and fat content.

Although the conservation by irradiation does not have any harmful effect on food consumption, an elementary ethics asks for a proper information of the public on any radiation treatment used for food decontamination.

On this subject it is worth mentioning that more countries, including the European Union, have issued a set of standard procedure to assess any irradiation treatment of a large category of food. In this regard, the electron spin resonance spectroscopy in view of its capacity to evidence minute amount of free radicals is the direct product of the action of high-energy ionizing radiation. For this purpose, the gamma-rays emitted by the ^{60}Co radioactive source are the most utilized. ESR spectroscopy is successfully used to assess the presence of free radicals in liquid media by means of spin traps and spin labels as well as directly, in hard part of food, where the free radicals are immobilized.

On the other hand, the ESR identification of irradiated food should be considered as a screening method: if ESR signal of radiation radicals is present, it can be supposed that the food was decontaminated by irradiation, and if this signal is absent, it cannot assert that the food was not irradiated.

Regardless of this small inconvenience, ESR analysis of food appears as an auspicious field of investigation, open to a great category of researchers.

Quiz

1. What are free radicals and how can their presence be evidenced by ESR?
2. Beside ESR, which are the other techniques which can be used to evidence the presence of free radicals in food?
3. Can thermoluminescence be used to assess the presence of a previous decontamination process by irradiation?
4. What are the differences between nitroso and nitrone spin trap?
5. Why does the nitroso ST give better information of the trapped free radicals?
6. What are the principal differences between spin traps and spin labels?
7. Why does the ESR investigation of irradiated food represent a screening procedure rather than a real assessment of any previous decontamination by high-energy ionizing radiation exposure?
8. Chitin is common for arthropods and mushrooms. What are the consequences of this fact for the food decontamination by irradiation?

Acknowledgments This work was supported by a grant of the Romanian Ministry of Research and Innovation, CCCDI UEFISCDI, project number PN-III-P1-1.2-PCCDI-2017-0686/52PCCDI/2018, within PNCDI III, by the Romanian Ministry of Research and Innovation, project number 15PFE/2018, and partially realized with the Cooperation Protocol No. 4322-4-2020/2022 between Joint Institute for Nuclear Research, Dubna, Russian Federation and University of Bucharest.

References

FAO (2020) http://www.fao.org/faostat/en/#home

IAEA (2008) Trends in Radiation Sterilization of Health Care Products, International Atomic Energy Agency, Vienna, 2008, ISBN: 978-9201110077

R.A. Mollins (Ed.) (2001) Food Irradiation: Principles and Applications, Wiley-Interscience, New York, ISBN:978-047135634

X. Fan, C.H. Sommers (Eds.) (2013) Food Irradiation: Research and Technology, Willey-Blackwell, and the Institute of Food Technologists, ISBN: 978-0813802091

M. Ikeya (1993) New Applications of Electron Spin Resonance, World Scientific Publishing, Singapore, ISBN 978-9810211998

G.A. Webb (Ed.) (2018) Modern Magnetic Resonance, Springer, ISBN: 978-3319283890

C.P. Pool, H.A. Farach (1994) Handbook of Electron Spin Resonance: Data Sources, Computer, Technology, Relaxation, and ENDOR, AIP Press, NY, ISBN: 978-0387986609

F.A. Willamena (2017) Reactive Species Detection in Biology: From Fluorescence to Electron Paramagnetic Resonance Spectroscopy, Elsevier, ISBN: 978-0124200173

L.M.L. Nollet, F. Toldr (Eds.) (2011) Safety Analysis of Foods of Animal Origin, CRC Press, Boca Raton, ISBN 978-1-4398-4817-3 419

A.G. Gaonkar (Ed.) (1995) Characterization of Food: Emerging Methods, Elsevier, ISBN: 978044481499X

A.M. Maghraby (Ed.) (2018) Topics From EPR Research, IntechOpen Ltd. London, ISBN: 978-1839620133

L.M.L. Noillet, F. Toldr (Eds.) (2015) Handbook of Food Analysis, CRC Boca Raton, ISBN 978-1466556546

S. Stoll, A. Schweiger (2006) EasySpin, a comprehensive software package for spectral simulation and analysis in EPR, Journal of Magnetic Resonance, 178, 42–55, https://doi.org/10.1016/j.jmr.2005.08.013

Index

© Springer Nature Switzerland AG 2021
A. K. Shukla, *ESR Spectroscopy for Life Science Applications: An Introduction*,
Techniques in Life Science and Biomedicine for the Non-Expert,
https://doi.org/10.1007/978-3-030-64198-6

Printed in the United States
by Baker & Taylor Publisher Services